The Implications of Induction

by the same author

THE DIVERSITY OF MEANING

THE PRINCIPLES OF WORLD CITIZENSHIP

The Implications of Induction

L. JONATHAN COHEN
Fellow of The Queen's College, Oxford

METHUEN & CO LTD
LONDON

First published 1970
by Methuen & Co Ltd
11 *New Fetter Lane, London EC*4

Printed in Great Britain
by Spottiswoode, Ballantyne & Co Ltd
London and Colchester

SBN 416 16000 X

Distributed in the U.S.A. by Barnes & Noble, Inc.

Contents

Preface

The first version of Chapter I of this book was read at a symposium of the British Society for the Philosophy of Science in September 1964. Later versions of Chapters I and VII constituted lectures to the University of Liége and the Centre National Belge de Recherches de Logique in March 1965; and reformulations of these lectures were published as 'What has Confirmation to do with Probabilities?', in *Mind* lxxv (1966), p. 463 ff., and 'A Logic for Evidential Support', in *Brit. Jour. Phil. Sci.* xvii (1966), pp. 21 ff. and 105 ff. But all these earlier versions contained errors of one sort or another, which I have tried to eliminate in the present book.

Similarly the ideas expounded in Chapters II to VI were first formulated in a number of lectures or papers that were delivered at about two dozen universities in Great Britain, Sweden and the U.S.A. during 1967–9.

I have profited a great deal from the questions, comments and criticisms that have been raised on these various occasions, though it is not possible to acknowledge them individually here.

During tenure of a visiting professorship at Columbia University in the autumn of 1967 I was fortunate to be able to discuss some of my ideas with Prof. E. Nagel, and I was greatly helped by this while writing the book. My interest in induction was first aroused by Prof. W. Kneale's lectures at Oxford in 1946–7, and I have also profited a great deal from his comments on a draft of the present book which he was kind enough to read. Prof. G. Ryle read the whole book in typescript, and I am very grateful for his many helpful comments on both style and content. Mr. J. L. Mackie and Mr. A. Margalit were also kind enough to read the typescript, and I learned a lot from their comments and criticisms and also from correspondence and discussion with Prof. I. Levi. Finally, my wife has nobly helped me to eliminate very many errors at various stages of the book's composition. But for the views expressed and the errors that remain I alone, of course, take full responsibility.

The writing of the book was greatly assisted by two periods of sabbatical leave that my College kindly gave me—during parts of the academic years 1963–4 and 1967–8.

L.J.C.

16 September 1969

Introduction

There is a certain way of writing about induction that seems at first sight attractively rigorous and professional. The writer begins by stipulating some rather exact criteria of adequacy for any proposed explication or analysis of inductive support. Next, he proposes his own explication of that concept. Finally he proves, with unimpugnable rigour, that his own proposal satisfies the stipulated criteria of adequacy (and perhaps even that no other proposal could possibly satisfy them). To take a very simple example: maybe it is stipulated that any function which maps one proposition's inductive support for another on to real numbers must satisfy both Kolmogorov's probability-axioms and also the requirement that, where a, b and c are distinct, Ra & Rb gives greater support to Rc than does Ra alone. Then, if a range-theoretical measure[1] like Carnap's c* is developed, its adequacy can be conclusively demonstrated. The uncertainties and imprecision of philosophical controversy have apparently been side-stepped, and a long-standing problem of epistemology has apparently been solved by a relatively elementary piece of mathematics.

But in fact no philosophical problem—no problem of analysis or explication—can ever be solved in this way. If criteria of adequacy are stated in such terms that their satisfaction or non-satisfaction in any particular case is capable of a purely mathematical proof, then they themselves are wide open to philosophical dispute. It cannot just be taken for granted, for instance, that inductive support is a Kolmogorovian probability. Barely to stipulate such criteria without defending them is a form of dogmatism that will not dispel any characteristically philosophical perplexity. But, if their defence is to be undertaken, it must take the form of philosophical argument; and, once the argument has been given, these logico-mathematical principles begin to appear as a kind of first-stage analytical conclusion, rather than as criteria of adequacy for the analysis as a whole.

So if we are to begin with adequacy criteria at all, in relation to the problem of inductive support, they must be philosophical rather than mathematical in nature. But how should they be chosen? An obviously

[1] By a 'range-theoretical measure' here I mean a measure of support that is a function of measures assigned to the ranges of sentences, where the range of a sentence s in a system L is the class of those state-descriptions in L in which s holds true.

unexceptionable policy is to select them in the light of the main varieties of argument that philosophers have offered, or could offer, against previous explications or analyses of inductive support. A new analysis must at least be able to surmount all the main difficulties that have floored the old ones.

I shall not, however, attempt in this book any detailed discussions of other philosophers' theories about induction. The recent literature on the subject contains many excellent critical discussions.[1] The need now is not for more criticism of old theories, but rather for new theories to criticise. In any case, a new theory stands out more clearly, and is easier to appraise, if it stands on its own and not as an outgrowth of older ones. So I shall begin here with just a bare list of what I take to be the main criteria of adequacy—adequacy of analysans to analysandum—that any new analysis of inductive support should aim to satisfy and every previous analysis has failed to satisfy completely. In later chapters each of these criteria will be discussed at greater length.

First, the analysis must apply to all the main types of proposition that are thought capable of support by the evidence of experimental science. It must apply to universal propositions as well as to singular ones, to propositions about causal connections as well as to non-causal propositions, and to quantitative correlations and scientific theories as well as to the hypotheses that are advanced at a more elementary level of scientific enquiry. Moreover, it must, if possible, treat inductive support in each case as something that admits of degree, not just as something that a hypothesis either has or has not.

Secondly, the analysis must not be hit by any of Hempel's or Goodman's paradoxes.

Thirdly, the analysis must apply to statistical patterns of scientific enquiry, as well as to deterministic ones.

Fourthly, the analysis must be consistent with the fact that scientists sometimes prefer to treat their hypotheses as being subject to qualification, rather than outright rejection, in the face of adverse evidence. An adequate analysis must trace the nature of the connection that exists between the extent to which evidence supports a hypothesis and the extent to which it enforces the latter's qualification; and it must also show how so-called 'fact-correcting' generalisations—idealisations that appear to tell us how nature should behave rather than how nature actually behaves—may be brought within the scope of inductive logic.

Fifthly, any logical analysis of inductive support must be in principle

[1] For references see H. E. Kyburg, 'Recent Work in Inductive Logic,' *Amer. Phil. Quart.* i (1964), p. 249 ff. Cf. also I. Lakatos, 'Changes in the Problem of Inductive Logic' in *The Problem of Inductive Logic*, ed. I. Lakatos (1968), p. 315 ff.

applicable to other subject-matters than those of experimental science. The empirical basis of scientific reasoning needs to be clearly distinguished from its logical structure.

Sixthly, the analysis must be more than just a description of some actually occurring patterns of argument. The patterns must be shown to be, in some relevant sense, rational and justifiable as well as actual.

Seventhly, and finally, the analysis must admit of deployment in an exact and systematic form, so that its implications are readily intelligible and its consistency is demonstrable. Axiomatic presentation is certainly not a sufficient virtue in itself. It is not a substitute for satisfaction of the other six criteria. But it is no less important than any of them.

The framework of the present book is determined by these seven criteria of adequacy. Chapter I is primarily concerned to elicit some consequences of insisting on close comparability between inductive support for a first-order generalisation and inductive support for its instances. One consequence is shown to be that support for any conjunction of such instances must normally be no more and no less than support for any single instance. A further consequence is that this support cannot be a function of any relevant logical probabilities, except in certain special circumstances. In Chapter II conjunctions of first-order generalisations are shown to obey substantially the same principles as conjunctions of their instances, if inductive support is conceived to depend on the results of experimental tests. Chapter III begins by showing how far this feature also characterises causal hypotheses and correlational generalisations, and it goes on to elucidate the role of what Whewell called 'consilience' in determining the level of support for scientific theories. The chapter also shows that, if inductive support is conceived to depend on the results of experimental tests in the way proposed, then neither Hempel's nor Goodman's paradoxes arise. Chapter IV is primarily concerned to show how the proposed analytical reconstruction of discourse about inductive support applies to statistical generalisations, and how argument by so-called enumerative induction, and argument from the results of experimental tests, are not to be regarded as co-species of a single logical genus. Chapter V shows, in effect, how any assessment of inductive support for a hypothesis on given evidence may be reinterpreted as an assessment of how far that evidence allows the hypothesis to be simplified. Once the possibility of such a reinterpretation is recognised, it is easier to discuss the operation of inductive patterns in legal reasoning from judicial precedent and in several other fields outside the scope of experimental science. Chapter VI distinguishes between the Cartesian type of scepticism, which requires reassurance, in the form of a philosophical proof, that the future will be like the past, and the Humeian type of scepticism which seeks a rational excuse, as it were, for our natural convictions that the future will resemble the past. It is then shown that, on a proper analysis of inductive support, syntactic analogies between induction and deduction suffice to rebut Hume's sceptical arguments and to legitimate the title of induction to be regarded as a mode of valid reasoning. Finally, in Chapter VII the logical syntax of inductive support, as argued for informally in preceding chapters, is mapped on to a class of generalised modal calculi, and inductive functors are shown to be definable in terms of the primitive symbolism of such a calculus. Inductive syntax, thus envisaged, is then developed systematically and in detail, and its consistency is established.

So the bulk of the book—the first six chapters—are philosophical in character, and formalisation is postponed to Chapter VII. No doubt one reader will find the earlier discussions too informal, and another will think the concluding chapter unnecessary. What I have tried to argue so far in this brief introduction is just that both such objections would be misconceived.

A reader of yet another type may find the whole discussion altogether too abstract, and too remote from the concrete complexities of actual science. Perhaps he thinks that every assertion about scientific reasoning needs to be fully documented with historical references, and he may complain that no actual reasoning has ever taken the form described in the following pages. But such a reader would be failing to apply to the philosophy of science a principle of method that has time and time again proved its value within science itself. This principle is that no real understanding, no potentially fruitful explanation, is ever achieved until the description and classification of concrete individuals, populations and substances gives way to the study of abstract properties and of the idealised individuals that are characterised by having or not having those properties. Natural history must give way to theory-construction within scientific enquiry itself, and within the philosophy of science an analogous transition has to be made. For example, we need to consider problems about the logic of science in abstraction from problems about its content, its methods of discovery, its economic organisation or its social function, even if these problems are to some extent theoretically interdependent and their manifestations are always in practice intertwined. Also, we need to consider idealised hypotheses and idealised reports of experimental tests, rather than actual ones, in order to be able to concern ourselves only with logically relevant properties. To complain just because this is done in a book on inductive logic is like complaining that there are no descriptions of a single actual billiard-game in a text-book of mechanics.

I am all too well aware that the argument of the book is long and complex. I have therefore prefaced each of the twenty-two sections by a summary of its contents. Those who seek a longer introduction than the present one may obtain a more detailed conspectus of the book from these summaries.

Some technical terms are inevitable, it should also be said, in a book about inductive logic. But their number can easily grow to a point at which they begin to obstruct philosophical objectives rather than assist them. I have therefore tried to operate with as few technical terms as possible, and to introduce new ones only where they have a substantial task to perform. For convenience of reference I have appended a short glossary of the latter.

I

Some Elementary Principles in the Logical Syntax of Experimental Support

§1. A characterisation of the topic

As a preliminary characterisation of the analysandum, consider the support given to a hypothesis by experimental evidence. But the domain of the support-relation to be studied is constituted by propositions, and the relation is a timeless one, unlike confirmation (in the ordinary sense). 'E supports H' is also to be distinguished, for several reasons, from 'E justifies accepting H'. It must be assumed that comparisons of inductive support are possible. But it will not be assumed at the outset either that inductive support is measurable or that it is not. Nor will it be assumed either that inductive support-functions take only empirical or scientific propositions for their arguments or that they do not take only these arguments.

The topic of this book is the logical syntax of inductive support. So to state precisely at the outset, in all relevant respects, what meaning should be attached to the phrase 'inductive support' would be to anticipate, oracularly, the book's intended conclusion. But for directing attention towards the conceptual field with which the book is concerned, and providing an initial premise from which to operate, it will suffice to remark that instances of at least one type of inductive support are familiar now to everyone and frequently described, viz. when a hypothesis about the solution of some problem in natural science has been tested experimentally and the report of this evidence is said to support the hypothesis.

If a man treats a proposition as a hypothesis he takes up a certain kind of attitude towards it. He is actively entertaining in his mind the possibility that it states the answer to his problem. Hence a hypothesis is not just a proposition, a unit of logical analysis. It is someone's working assumption. Infinitely many propositions are never hypothesised at all, even if alike in logical structure to those that are. But just as truth-functional logic and quantification theory are not restricted in application to propositions that are, or should be, actually asserted in the course of human history, so too the logical syntax of inductive support is best studied without restriction to the consideration of support for propositions that are, or should be, actually hypothesised at some particular time and place. The concept of a hypothesis may have an important role to play in the history and heuristics of natural science. But it is not intrinsic to the logic of inductive support.

Similarly, when a proposition is treated as the report of an experimental test, the facts stated in the proposition are supposed to be due, in part, to deliberate human contrivance—in the planning of the test, selection of materials, control of relevant variables, construction of any necessary apparatus, and so on. But such inductive support as this proposition may give to another is not dependent on the facts being thus due to human contrivance. Rather, when an experiment is, or should be, contrived, it is, or should be, contrived in order to discover whether a proposition that in any case gives this support is actually true. The concept of an experimental test, like that of a hypothesis, does not belong to the logic of inductive support, but presupposes it—in the sense that there is no point in carrying out an experimental test unless such or such a support-relation exists between the propositions concerned.

In short, when one cites as an instance of inductive support the support-relation between the report of a series of experiments and the hypothesis it was designed to test, this relation is to be understood as a relation between two propositions E and H that holds good irrespectively of the part that is or ought to be played by human beings in bringing about the facts stated in E and irrespectively of the attitude that is, or ought to be, taken up by human beings towards H.[1] Hence, though the verb 'confirms' is often used in place of, or as a synonym for, 'supports' here, it is potentially misleading when so used, as by Carnap.[2] What we normally take to be open to confirmation are human commitments, such as hotel reservations, administrative decisions, scientists' hypotheses, or astrologers' predictions. Once the prediction has been advanced, the hypothesis entertained, the decision tentatively taken, or the reservation provisionally booked, it is thereafter open to confirmation by the appropriate events or actions. Thus normally, if x confirms y, then x is later than y, and both x and y must be datable. But propositions, as distinct

[1] In what follows I nevertheless often write, loosely, of the support given to a hypothesis by the results of experimental tests. The reader should treat such phraseology in the text as a stylistic device for avoiding the cumbersome periphrases and repetitions that would be necessary if I were to speak always of propositions. Cf. p. 116, n. 1. *Stricto sensu* 'hypothesis' is a term in the theory of acceptability (cf. p. 8 ff.), not in that of support.

[2] R. Carnap, *Logical Foundations of Probability*, 2nd ed. (1962), p. 1 f. and *passim*. Since 'to confirm' means, literally, to strengthen, i.e., to make something stronger than it was before, the verb sheds temporal implication with difficulty. Admittedly there is a spatial metaphor dormant in the philosophical use of the verb 'to support'. But a spatial metaphor does not have the same opportunity to give trouble in the present context as a temporal one.

from their utterances, are not datable.[1] So to speak of inductive confirmation may tend to direct attention away from the logical relation of support towards consideration of some historically or heuristically important relation that presupposes the logical one but is not co-extensive with it. Science may or may not have made greater progress through the work of people who first entertained hypotheses and then proceeded to test them, than through the work of people who first catalogued facts and then tried to formulate appropriate generalisations. But this issue is necessarily quite irrelevant to consideration of inductive support as a relation between propositions, if propositions are not datable. So where the topic to be explored is the logic or logical syntax of inductive support, not the methodology of scientific discovery, the word 'confirmation', as a term of ordinary language, cannot usefully serve as a preliminary indication of subject-matter, whatever its later utility as a technical term in some developed philosophical theory. Carnap's problem is not the same as Popper's,[2] though Carnap's use of the word 'confirmation' helps to suggest that it is.

Nor is it to be assumed too readily that the solution of a given problem which has the most inductive support from already known truths is necessarily the best solution of that problem at the moment. The question of acceptability, in the sense of acceptance-worthiness—the question of which hypothesis constitutes the best solution of a given scientific problem—needs to be carefully distinguished from the question of inductive support. There are at least three important differences between these two issues, quite apart from the fact that acceptability is relative to a problem in a way that support is not.

First, the extent of E's inductive support for H is independent of whether or not E states all the available evidence about H. Indeed a large part of the point of discussing whether or not E comprehends all the available evidence about H, when E is at least true, springs from the possibility that some other proposition F, logically implying E but supporting H more than E does, or supporting some other solution more than E supports H, may also be true. One must thus be able to assess the extent of E's inductive support for H without making assumptions about E's comprehensiveness, in order to be able to decide whether or not it is worth while to investigate the truth of F. But the situation in regard to acceptance is quite different. So far as H's acceptability as the answer to a given question is to be judged by an assessment of inductive support,

[1] By 'proposition' here I mean what I have elsewhere called a statement: cf. *The Diversity of Meaning*, 2nd ed. (1966), p. 163 ff.

[2] Cf. K. R. Popper, *Conjectures and Refutations* (1963), p. 47 ff., and *The Logic of Scientific Discovery* (1959), p. 251.

the support must be that afforded by a proposition which is assumed to state all available relevant evidence. To decide about acceptance on the basis of a partisan selection from the evidence is obviously improper: almost any solution of a problem could be made to appear acceptable that way. The difference here between inductive support and inductive acceptability may be illuminated by comparison with the corresponding difference between the question whether one proposition is deducible from another and the question whether this proposition is the best answer to a given question that is deducible from known or postulated premises. Such-and-such a theorem may perhaps be provable in a system, on the basis of some of the system's postulates, but if all the postulates are used a more general or more elegant theorem may be obtainable. *Pace* Carnap,[1] therefore, the requirement of total evidence has an important analogue in relation to deducibility. But this cannot be seen until the requirement is acknowledged to be relevant only to questions about acceptability—about the best solution of a given problem—not to questions about support.

It follows, secondly, that judgements of acceptability are relative to time, place and circumstances of various kinds, where these affect the availability of inductive evidence.[2] A solution to some chemo-therapeutic problem, that deserved acceptance last year in the light of what was then known about the properties of the drugs involved, may come to deserve rejection later on, when hitherto unsuspected side-effects of the drugs begin to turn up in special cases. Moreover, evidence may be out of a man's reach not only because it does not yet exist, but also because it has not yet been published or is too expensive to obtain, or because no-one will ever be sufficiently ingenious to devise the experiments that would produce it. Even if one excludes laziness and dishonesty by requiring, like Popper, that a hypothesis should be tested as severely as possible, this will still leave it open to draw the line between evidential availability and unavailability in a very wide variety of ways. How long has a scientist to ponder before being entitled to conclude that no further experiments can be thought up to test his hypothesis? How hard must he press the sources of finance before being entitled to conclude that funds for such-or-such an experiment will not be forthcoming? How much must he read before being entitled to conclude that no other scientist's work supports, or undermines, his ideas? But judgements of inductive

[1] *Logical Foundations of Probability*, p. 211.

[2] It is important to distinguish here the conditions of inductive detachment, which is a purely logical operation and holds good—if it holds good at all—irrespective of what human beings happen to know at any one time, from the conditions of inductive acceptability, which relate closely to available knowledge: see p. 68 ff. below.

support, unlike judgements of acceptability, are not relative to time, place or circumstances, because they are independent of evidential availability. A man may feel he knows just how much a certain experimental result would support his hypothesis, even if the actual performance of the experiment is quite outside his power. Admittedly, his views about how much E supports H may change, and rightly change, as he comes to see that certain new tests on hypotheses of a given type are superior to the old ones. The history of pharmacology is full of such changes. But that is no reason for supposing that inductive support itself is relative to a time, let alone to a place or to other circumstances. Rather, the discovery that one is ever wrong about the extent of E's support for H is going to be very difficult indeed if this extent itself can alter. Just as the propositions related by inductive support have no spatio-temporal locations, so too the relation of inductive support is independent of time and place. Just as the experiments alleged to favour a particular hypothesis are, in principle at least, indefinitely repeatable, so too the extent of the support that their description affords the hypothesis remains the same whenever and wherever the experiments are performed. Otherwise what would be the point of claiming that these experiments are indefinitely repeatable?

A third difference between judgements about acceptance and judgements of inductive support is that the former may legitimately appeal to considerations which are quite irrelevant for the latter. Suppose a pharmacologist's problem is to determine whether a certain drug is free from toxic side-effects. An affirmative answer to this problem may deserve acceptance on less supporting evidence if the disease to be treated by the drug is often fatal than if it never is. In the light of such cases some philosophers have claimed, like Braithwaite,[1] that ethical issues are bound to arise within inductive logic. But it will be assumed here that such issues are always as irrelevant to judgements of inductive support as they are to judgements of deducibility. Where ethical issues are relevant to judgements about hypotheses is in relation to acceptance. To the extent that we accept such-or-such a hypothesis as the solution of a given problem, we are ready to act on that hypothesis where action is necessary, and so it is reasonable to take into account the consequences of so acting when we decide under what conditions a particular kind of hypothesis deserves acceptance. Nor is this difference between judgements of acceptability and judgements of inductive support confined to hypotheses that have a direct practical application. The advantage gained by accepting one hypothesis as against another, at least for a while, may be theoretical rather than practical. It may open

[1] R. B. Braithwaite, *Scientific Explanation* (1953), p. 174.

up a new field of possible investigation—a range of experimentally answerable questions that would not otherwise be asked—or achieve a wider explanatory synthesis. To that extent a more general, or less qualified (i.e. simpler), hypothesis may be more acceptable, even if less well supported. In short, though a judgement of inductive support must be one premiss of any argument for acceptability, it is not the only possible kind of premiss for such an argument.

No doubt in the concrete contexts of life questions of acceptability are the questions that scientists, engineers, physicians, etc. need ultimately to answer. But the study of complex questions can nearly always profit by our considering some of their components in abstraction from one another, and a separate investigation into the logical syntax of inductive support is not to be condemned just because it cannot provide the whole answer to any question about acceptance.

To the older writers, like Bacon and Mill, the distinction between inductive support, on the one side, and acceptability, on the other, was not apparent. There were two main reasons for this. First, they normally treated inductive support as a matter of all-or-nothing rather than of more-or-less. A hypothesis was to be classified either as established or as not established. So no problem could arise for them about the level of inductive support that justifies accepting a hypothesis of such-or-such a type. Secondly, they tended to think of induction both as a method of discovery and as a form of support or justification; and if induction is the heuristics of causes, as Bacon and Mill often viewed it, it typically reveals that certain hypotheses deserve acceptance.

But there is no room for a purely qualitative or classificatory concept of inductive support, once questions of acceptance are distinguished from questions of support. A principal merit of this distinction is that it enables us to talk about the *level* of support by available evidence at which a hypothesis deserves acceptance. Admittedly some modern philosophers, like Nicod,[1] have discussed criteria for calling an event a 'supporting' or 'confirmatory' one in relation to a given hypothesis. But this discussion need not invoke a qualitative or classificatory concept like Mill's and Bacon's. Such philosophers may be construed as seeking criteria for saying that the support given by E to H is of more than zero extent, or more than that given typically by a tautology. It will be assumed at the outset here, therefore, that inductive support is at least a comparative relation, if not also a rankable or measurable one. Not that we can, in contemporary science, compare E's support for H with F's for I in the case of every E, H, F and I. But within particular fields of scientific research such comparisons are commonly thought reputable

[1] J. Nicod, *Foundations of Geometry and Induction* (1930), p. 219.

and afford at least part of the basis necessary for judgements about acceptance.

Some writers have denied that it is possible to systematise these comparisons into a theory that treats inductive support as a measurable quantity and still has adequate backing from the practices and intuitions of scientists. But in any case we can at least explore the conditions that any concept of measurable inductive support must satisfy. Reputable comparative judgements about inductive support are indisputably possible, and some of the conditions bearing on any possible method of measuring inductive support must be determinable by reference to the conditions bearing on comparative judgements. For example, if I's deducibility from H guarantees the truth of the comparative judgement $s[H,E] \leqslant s[I,E]$—i.e. the judgement that E's support for I is not less than E's support for H—then where I is deducible from H any quantitative theory must require such values to be assigned to s[H, E] and s[I, E] as will maintain this inequality invariant. We shall certainly be in a better position to judge whether or not any proposed measure of inductive support is satisfactory when we know as much as the study of comparative judgements can tell us about the logical framework within which any quantitative measure must operate. Moreover, even if no scale of measurement is conceivable whereby a support-function s[. . ., ---] will take real numbers as values, it may nevertheless be possible to descry plausible criteria for ranking or grading support-relations, whereby s[. . ., ---] will take finite ordinals as values.

Thus the concept of inductive support with which this book is concerned is one that distinguishes inductive support both from confirmation and from the justification of acceptance. It is a concept that admits of comparative judgements, and, at least *prima facie*, does not exclude the possibility of ranking or measurement. It is familiarly exemplified, as the concept of a certain timeless relation between propositions, in assessments of the support that experimental evidence gives to scientific hypotheses. But one would be wrong to suppose at the outset of the enquiry that the concept of inductive support is applicable only in this kind of context or in everyday approximations to it. Certainly no such supposition is normally made by formal logicians about deducibility. Every proposition, irrespective of its content, is capable of having deductive support. A proposition is logically true if it remains true under all uniform replacements of its non-logical terms,[1] and so a proposition asserting the deducibility of H from E remains true when any scientific

[1] This is to be regarded as a criterion of logical truth in unformalised discourse. It is sometimes objected against such a criterion that there is a certain type of existential proposition which is logically true according to the criterion but would not normally be

or empirically descriptive terms in H and E are uniformly replaced by mathematical, theological or ethical ones. There is thus some justification, by analogy with deductive logic, for suspecting that, despite the preoccupations of men like Mill, Popper and Carnap, the domain of inductive support is not restricted to propositions about the subject-matter of empirical science. Hence, just as in relation to the problem about a quantitative measure it is worth while determining, so far as is possible, from the relatively uncontroversial study of comparative judgements, what general conditions such a measure must satisfy, so too in relation to reasoning about non-empirical subject-matters it is worth while determining, so far as is possible, from the relatively uncontroversial study of experimental support for scientific hypotheses, what general conditions such reasoning must satisfy if it is to be classed as inductive. In both cases an enquiry into the logical syntax of experimental support should disclose logically necessary, though not—for the most part—logically sufficient, conditions for the assignment of values to inductive functions. It is quite another question, with which this book is somewhat less concerned, to ask systematically what sufficient conditions, if any, can also be stated for particular types of value-assignment. The main object of study is the logical syntax of inductive support-functors,[1] not their semantics.

§2. A derivation of the instantial conjunction principle

The logical syntax of inductive support-functors will be taken to define a general class of inductive functions, a sub-class of which is constituted by the support-functions invoked in experimental science, just as the mathematical calculus of probabilities may be taken to define a general class of syntactically similar, though semantically dissimilar, probability-functions. But can the logical

regarded as logically true. Specifically, if there are n things in the universe, the criterion seems to make

(A) $\exists x_1 \exists x_2 \ldots \exists x_n(x_1 \neq x_2 \,\&\, \ldots \,\&\, x_i \neq x_j \,\&\, \ldots \,\&\, x_{n-1} \neq x_n)$

logically true, whereas the assertion that just so-and-so many things exist would normally be regarded as contingent. But this objection is fallacious. We cannot count mere things or entities, but only chairs, apples, atoms, prime numbers, etc.; and, even if we choose chairs for our domain of discourse, 'There are just n chairs in the universe' is not logically true by the proposed criterion, since we could have chosen a different domain. (A) has the appearance of a counter-example, only because in reading a formula of quantification theory we often assume a specific universe of discourse tacitly and the assumption does not then appear to have been made at all. Hence the only purpose really served by the objection is to remind us that when we speak of a proposition (as distinct from a propositional schema) we imply a definite domain of discourse. A proposition is logically true in unformalised discourse if it remains true under all uniform and grammatical replacements of the non-logical terms in any sentence that expresses it, *including those that specify, implicitly or explicitly, its universe of discourse.*

[1] I use the term 'functor' for the notational devices that express functions.

syntax of inductive functors be mapped on to the probability-calculus in any way? In order to provide a basis for answering this question two important syntactical principles will be assumed, viz. that the inductive functor is open to substitution on the basis of non-contingent equivalence (equivalence principle), and that the comparative value of s[U, E] normally varies with the comparative value of s[P, E] when U is a universal proposition of which P is a substitution-instance (instantial comparability principle). From these two assumptions it is possible to prove that normally $s[P_i, E] = s[P_1 \,\&\, P_2 \,\&\ldots\&\, P_n, E]$ where $P_1, P_2, \ldots$ and P_n are all substitution-instances of the same universal proposition (instantial conjunction principle). If P_1, P_2, etc. are propositions about causal connections, it is possible to derive both the instantial comparability principle and the instantial conjunction principle from the equivalence principle alone.

The distinction (drawn at the end of the preceding section) between the syntax and semantics of inductive support-functors may be illuminated by comparison with the corresponding distinction in the case of probability-functions.

At one time philosophers used to assume, in effect, that the mathematical calculus of probabilities is not open to more than one correct interpretation. In this monistic atmosphere Laplace's recourse to the principle of indifference, the frequency interpretations of von Mises, Reichenbach, etc., and the logical probabilities of Maynard Keynes, Harold Jeffreys, etc. were pitted against one another in apparently irresoluble conflict. But now it is widely accepted—for example, in Carnap's discussion[1] of what he calls probability$_1$ and probability$_2$—that the calculus may have more than one interpretation as a theory of probabilities. Indeed the calculus seems interpretable at least as a theory of the ratio, whatever actually happens, of the favourable to the equally possible chances, or as a theory of the actual relative frequency of a certain class of occurrences within a certain other class, or as a theory about range-overlap between propositions, and no doubt in many other ways too. What the formal system tells us, for the most part, is not how to determine probabilities from scratch, as it were, but in what ways the values of a certain functor of one or two elements restrict one another, where these values are all real numbers in the closed interval (1, 0). Only in certain limiting cases, such as $p[X, X] = 1$, does the formal system itself assign a value to the functor. For the most part it is concerned with universally valid compatibilities and incompatibilities between possible value-assignments: the multiplication-law for the probability of a

[1] *Logical Foundations of Probability*, p. 23 ff.

conjunction, the addition-law for that of a disjunction, and so on. Nor does the calculus itself specify the domain of the functor. The latter may take names of sets, of propositions or of anything else as fillers for its argument-places, so long as there is a function of these that satisfies the formal laws. In short, the calculus gives us a logical syntax for the probability-functor, not a semantics. Hence its axioms can conveniently be taken as constituting a definition for the generalised concept of a probability. Whatever the word 'probability' means, or has meant, in the actual discourse of scientists and others, we can now decide to say that two semantically heterogeneous functions are both to be regarded as probabilities if, and only if, they share this same formal syntax, though some such functions may, for historical reasons, be regarded as non-standard interpretations of the symbolism. Indeed in the present age it is bound to create unnecessary verbal confusion if philosophers seek to talk about probability in a sense of that word which does *not* imply conformity to the axioms of a mathematical theory that is known familiarly and universally as the calculus of probabilities.

It is fruitful to conceive a calculus of inductive support analogously. On the one hand there is the semantical problem of deciding, in each kind of case, what are the fillers for the argument-places of our inductive functor, and what criteria are most appropriate for deciding questions of equality or inequality or even perhaps for assigning ordinal or cardinal numbers as values of the functor for given fillers of its argument-places. On the other hand there is the syntactical problem of determining any compatibilities or incompatibilities that hold universally between such assignments or between such an assignment and the formal-logical properties (e.g. tautologousness) of the elements involved. To construct a calculus of inductive support is to solve the latter problem, not the former. Moreover we must bear in mind the possibility that such a calculus may admit of one or more non-standard interpretations. That is, by constructing a separate theory of the syntax of the inductive support-functor, as distinct from its semantics, we cannot but achieve at the same time a definition for a syntactically generalised concept of inductive function. Any function that satisfies the axioms of such a calculus will be an inductive function, even if its semantically appointed domain embraces non-empirical propositions and even if it is customarily regarded as an index of, say, simplicity or explanatory power rather than of inductive support. But just as the theory that gives the logical syntax of the probability-functor is not a *wholly* uninterpreted calculus or logistic system, so too, it should be noted, a calculus of inductive support is not a mere logistic system. In the former case one postulates that numerical expressions like '1' and '0', and arithmetical signs like '=',

‘>’, ‘+’, ‘×’, etc. always have their standard interpretation, even though various interpretations are possible for the rest of the symbolism; and the appropriate corresponding postulates—whatever they may turn out to be—will have to be set up for a theory about the logical syntax of inductive support.

There may be a sense of ‘inductive’ in which it is true to say that more than one syntax of inductive support is possible, as well as more than one semantics, and correspondingly we may need more than one calculus—or type of calculus. But there comes a level of abstraction at which no relevant positive reasons can any longer be found for grouping two sets of things under the same concept. To use the same word as the name for both is then either to rely on merely negative similarities or to take advantage of a homonym, the etymological origins of which afford no philosophically adequate ground for asserting conceptual identity. In this book therefore it will be assumed that, though very many semantically different inductive functions may fall under one and the same syntactical definition—some of them certainly being support-functions—nevertheless no two functions should both be called ‘inductive’ if their logical syntax is substantially disparate. Just as it is convenient to have only one definition for the concept of a probability-function, though there may be many such functions, so too it is convenient to have only one concept of an inductive function, even though there may be more than one function of this kind also. If there are functions of other kinds that have in the past been called ‘inductive’, or are closely associated in some way with statements about inductive support, it will be convenient to find another name for them. This will not diminish their importance or the sweetness of their smell.

Many philosophers have, in effect, identified a calculus of inductive support with the calculus of probabilities. The most direct way of doing this is to hold that inductive support-functions have precisely the syntax of probability-functions. Thus Reichenbach argued[1] that they measure the relative frequency with which entities of certain kinds occur among entities of certain other kinds; and Carnap[2] has treated E’s inductive support for H as a special kind of probability that is given by the ratio of the range-measure of E & H to the range-measure of E, where the range of a proposition in a language *L* is the class of state-descriptions in *L* in which the proposition holds true. But there are also many indirect ways of identifying a calculus of inductive support with the calculus of probabilities. Philosophers have often treated inductive support as being

[1] H. Reichenbach, *The Theory of Probability*, tr. E. H. Hutten and M. Reichenbach, 2nd ed. (1949), p. 434 ff.

[2] *Logical Foundations of Probability*, p. 293 ff.

a function of certain probabilities rather than a probability itself. For example, Good has claimed[1] that E's support for H, given G, is a logarithm of

$$\frac{p[E, G \& H]}{p[E, G \& \bar{H}]}$$

while Popper argued[2] that if E reports the severest tests devisable against H, and H is not falsified by them, then E's degree of support for H is identical with the logical improbability of H or with the logarithm_2 of this improbability.

There are therefore three distinct levels at which probabilistic theories of inductive support may be attacked. First, it is possible for a philosopher to criticise a particular probabilistic theory about the semantics of the inductive support-functor while accepting that it has just the same syntax as the probability-functor. Carnap's opposition to Reichenbach's theory[3] is an example of criticism at this level. Secondly, and more radically, it is possible to criticise the thesis that the inductive support-functor, has just the same logical syntax as the probability-functor, while accepting that the syntax of the inductive support-functor may be mapped on to the calculus of probabilities in some other way. Some of Popper's arguments[4] against Carnap are examples of anti-probabilism at this level. Thirdly, it is possible to criticise the thesis that there is at least some way in which the syntax of the inductive support-functor may be mapped on to the calculus of probabilities. This is the most radical level of attack on probabilistic theories, and it is the one that will be pursued in the present book. It is directed just as much against Popper's type of theory as against Carnap's. Not that the assessment of probabilities is irrelevant to the evaluation of experiments. Far from it. But its relevance will turn out (see §§12–14 below) to be to other kinds of evaluation than to judgements of inductive support.

As an initial basis for this very radical attack on probabilistic theories two major assumptions are necessary. The first is that the values of inductive support-functions remain invariant when their arguments are replaced by logically or mathematically equivalent propositions, i.e.

[1] I. J. Good, 'The White Shoe is a Red Herring', *Brit. Jour. Phil. Sci.* xvii (1967), p. 322 ff.

[2] *The Logic of Scientific Discovery*, pp. 387–419.

[3] *Logical Foundations of Probability*, p. 175 f.

[4] *The Logic of ScientificDiscovery*, p. 390 ff. These arguments, however, are invalid: cf. L. Jonathan Cohen, 'What has Confirmation to do with Probabilities?', *Mind* lxxv (1966), p. 463 ff. Popper does not use the word 'inductive' to describe his corroboration-functor, but he nevertheless claims that it expresses 'the intuitive idea of degree of support by empirical evidence' (*The Logic of Scientific Discovery*, p. 393, cf. pp. 395–6, 399, 410), which is what I have taken to be the analysandum for what I am calling a theory of inductive support.

(1) For any propositions E, F, H and I, if E is equivalent to F, and H to I, according to some non-contingent assumptions, such as laws of logic or mathematics, then s[H, E] = s[I, F].

Let us call (1) 'the equivalence principle for the logical syntax of dyadic[1] inductive support-functors' or, for short, 'the equivalence principle'.

It is very difficult to think why anyone should want to contest this principle in relation to the evidential argument-place, i.e. the requirement that if E is logically or mathematically equivalent to F then s[H, E] = s[H, F]. After all we are mainly interested in evidential propositions so far as their truth is capable of establishment by experiment or some analogous process, and we certainly expect that, if E is formal-logically or mathematically deducible from F and F from E, then the truth of E is established experimentally if and only if the truth of F is so established.

But there is at least *prima facie* plausibility in the suggestion, which is sometimes advanced, that in order to solve Hempel's paradoxes of confirmation[2] we should reject the equivalence principle in relation to the hypothesis argument-place, i.e. the requirement that if H is logically or mathematically equivalent to I then s[H, E] = s[I, F]. For, if we reject this requirement, we need not suppose—as it is indeed paradoxical to suppose—that the report of a non-black non-raven provides us with as good inductive support for 'All ravens are black' as for its contrapositive equivalent 'All things that are not black are not ravens'. But such a solution of Hempel's paradoxes is bought at a very high cost. If we suppose that the inductive support-functor resists substitution of logical equivalents for one another in its first argument-place, we assign that place to an even higher grade of non-extensionality—to a murkier type of referential opacity—than any to which we customarily assign the first argument-place in classificatory assertions that such-or-such a proposition is a law of mathematics or formal logic. If assertions that such-or-such a proposition is a law of natural science are more like propositions of the latter type, than they are like indirect discourse about what people have actually said or about the content of their minds, the equivalence principle ought to hold good. We certainly cannot be sure that a man who says, or entertains the thought, that H is true also says or entertains the thought that I is true, even if H and I are logically equivalent to one another, since this logical equivalence may not be evident to him. But the logical liaisons of an inductive support-functor will be very restricted indeed if we reject the equivalence principle in relation to its hypothesis argument-place.

[1] On the difference between monadic and dyadic support-functions cf. §8 below.

[2] C. G. Hempel, 'Studies in the Logic of Confirmation', *Mind* liv (1945), p. 1 ff. and p. 97 ff., reprinted in C. G. Hempel, *Aspects of Scientific Explanation* (1965), p. 3 ff.

Indeed, we must then also reject or restrict the consequence principle for hypotheses (for dyadic support-functions), viz.

(2) For any E, H, and I, if I is a consequence of H according to some non-contingent assumptions, such as laws of logic or mathematics, then $s[H, E] \leqslant s[I, E]$,

since in relation to the hypothesis argument-place (1) is logically deducible from (2). But how far can we afford to restrict (2)? Certainly we should have to retain the thesis that inductive support is passed on by a universal proposition to its substitution-instances. There would be little point in scientific generalisation if the applications of a universal hypothesis U to particular cases were not at least as well supported as U itself by the evidence for U. Our general knowledge would be useless. But just the same would be true of scientific system-building—the axiomatisation of a body of universal hypotheses—if the consequence principle for hypotheses were not accepted in its usual, unrestricted form. There would be little point in such systematisations if evidential support were not automatically passed on by a conjunction of universal propositions to any other universal proposition logically deducible from it. So it seems that (2) cannot be safely restricted in such a way as to invalidate the equivalence principle for hypotheses, and that any comparative, ordinal or quantitative concept of inductive support which failed to satisfy the latter principle would be of relatively little use or importance. Hempel's paradoxes of confirmation are therefore better resolved in some other way than by the rejection of (1), and an alternative solution will in fact emerge below.[1]

The other assumption to be made, alongside (1), will be called 'the instantial comparability principle (for dyadic support-functions)'. According to this assumption not only does inequality of inductive support for two generalisations lead normally to inequality of inductive support for the predictive or retrodictive applications of these generalisations to individual cases, but also an inequality of the latter kind normally requires an inequality of the former kind. More precisely

(3) For any E, E′, P, P′, U and U′, if U and U′ are first-order generalisations, P and P′ are just substitution-instances of U and U′ respectively, and E and E′ are conjunctions of existenti-

[1] See §11. It is no doubt also possible to argue here that any probabilist theory of inductive syntax is committed to the equivalence principle and therefore that any attack on probabilist syntax is entitled to assume it. But to invoke such an *ad hominem* argument at this point is to suggest that interest in the equivalence principle may be confined to its destructive role in providing a basis for the criticism of probabilist syntax, whereas in fact the equivalence principle belongs to the logical syntax of inductive support in its own right and therefore needs a defence that is independent of the argument against probabilism.

ally quantified propositions each of which predicates something of some unspecified individual element in the domain of discourse, then $s[U,E] > s[U',E']$ if and only if $s[P,E] > s[P',E']$,

where by 'just a substitution-instance' is meant that all the relevant information given by P and P′ is that the elements to which they refer satisfy U and U′ respectively: their modes of referring to these elements add no further relevant information.

If we come to make new tests on the basis of which we conclude a scientific hypothesis to be even better supported by evidence than we had hitherto supposed, we would normally infer a correspondingly increased measure of support for each singular prediction or retrodiction we derive from that hypothesis. Similarly, if a man is told that a singular prediction —about, say, the curative value of the pills in *this* box for *his* illness—is better supported by the evidence now available than by the evidence previously available, where neither lot of evidence says anything about the man or box in question, the man would normally suppose that the evidence now available lends greater support to some general hypothesis about the curative value of *that kind* of pill for *that kind* of illness. Admittedly, if the fact described by P′ were included in the evidence, E′, for U′, while the fact described by P was not included in the evidence, E, for U, where P′ and P are substitution-instances of U′ and U, respectively, one might then well expect to find that $s[U,E] > s[U',E']$ was compatible with $s[P,E] \leqslant s[P',E']$, whereas the consequent of (3) requires $s[P,E] > s[P',E']$. But (3)'s antecedent excludes such a situation by requiring that E′ be a conjunction of existentially quantified propositions each of which predicates something of some unspecified element in the domain of U′.

It is difficult to believe that anyone can refuse to accept (3) unless he has already adopted a philosophical theory of inductive support whereby he has become deeply committed to an opposing viewpoint.[1] The con-

[1] Certain logically false propositions, e.g. those of the form $(x)(Rx \mathbin{\&} -(y)Ry)$, constitute apparent counter-examples to (3). Let U be such a proposition and P be some substitution-instance of it, $Ra \mathbin{\&} -(y)Ry$, and let E state that a large number of different things are R and just a few are not R. Then it would seem that U, being logically false, would have no support from E, while P might have quite a lot of support. So there might well be a U′ and a P′ in (3) such that U′ was better supported than U even though P was better supported than P′. One way to deal with this situation is to interpret it as showing the need to qualify (3) by requiring both U and U′ to be logically consistent. Another way is to interpret the situation as showing that, whatever be the criteria of support according to which such an E may be said to support such a P, these criteria are not appropriate to a concept of support that satisfies (3). It turns out that other considerations favour the latter policy: $(\exists y) - Ry$ does not inductively support, though it logically implies, $-(y)Ry$—cf. §8, p. 61, and §20, p. 199, below. It is also easy to find apparent counter-examples to (3) if one neglects the proviso that P and P′ are *just* substitution-instances of

nection that (3) asserts between inductive support for generalisations and inductive support for their instances seems an essential condition for our being able to make useful comparative judgements about the former and well-grounded ones about the latter. Moreover, though this instantial comparability principle has been curiously overlooked hitherto, it has at least one very important logical consequence when conjoined with the equivalence principle. Any inductive function that satisfies (1) and (3) must also satisfy the principle that any conjunction of instances of a generalisation normally has the same support on given evidence as a single instance. More precisely, this instantial conjunction principle for dyadic support-functions asserts that

(4) For any E, P_1, P_2, ..., P_n and U, if U is a first-order universal proposition with a domain of at least *n* elements, and P_1, P_2, ..., and P_n are just some substitution-instances of U, and E is a conjunction of existentially quantified propositions each of which predicates something of some unspecified individual element in the domain of U, then $s[P_i, E] = s[P_1 \& P_2 \& \ldots \& P_n, E]$.

In order to derive (4) from (1) and (3) it is first necessary to derive a corollary of (3), viz. the principle that in the circumstances described by (3) we have

(5) If $s[U, E] = s[U', E']$, then $s[P, E] = s[P', E']$.

This derivation is easily achieved since in those circumstances we shall have, by contraposition of each of the conditionals asserted in the consequent of (3), both

If $s[U, E] \leqslant s[U', E']$, then $s[P, E] \leqslant s[P', E']$,

and also

If $s[U, E] \geqslant s[U', E']$, then $s[P, E] \geqslant s[P', E']$.

We shall therefore have

If $s[U, E] \leqslant s[U', E']$ and $s[U, E] \geqslant s[U', E']$,

then $s[P, E] \leqslant s[P', E']$ and $s[P, E] \geqslant s[P', E]$,

which gives us (5).

U and U′, respectively. A. C. Michalos, in 'An Alleged Condition of Evidential Support', *Mind* lxxviii (1969), p. 440 f., produced three such examples in valid criticism of an earlier version of the instantial comparability principle that appeared in my 'What has Confirmation to do with Probabilities?', *Mind* lxxv (1966), p. 469 f. But such examples do not hit the slightly later version that appeared in my 'A Logic for Evidential Support', *Brit. Jour. Phil. Sci.* xvii (1966), p. 23, or the still later version that appears in the present book.

Next, it is to be noted that, if U^1 is a universal proposition with a domain of at least n elements, then certain other universal propositions U^2, U^3, ..., U^n are logically equivalent to U^1. Consider first, for simplicity's sake, the case where only one universal quantifier occurs in U^1. Then these equivalents are formed by generalising analogously not about single elements of U^1's domain, but about at most two, three, or n of these elements. For example, 'All ravens are black' has as one of its equivalents 'For all x and y, the same or different, x is a raven only if black and y is a raven only if black'. Moreover, if U^i is such an equivalent of U^1, the equivalence principle (1) gives us for any E

(6) $s[U^1, E] = s[U^i, E]$.

It follows that in the circumstances described by the instantial comparability principle (3) we are entitled by (5) and (6) to assert

(7) $s[P^1, E] = s[P^i, E]$

where P^1 is a substitution-instance of U^1, and P^i of U^i. For example, E supports the prediction that a is a raven only if black to just the same extent as E supports the prediction

(8) a is a raven only if black and b is a raven only if black.

Now let P_1, P_2, ... and P_n be the propositions about individuals that correspond to (i.e. are conjoined in) P^i in the way that, e.g. 'a is a raven only if black' and 'b is a raven only if black' corresponded to (8). Then (7) may be rewritten

(9) $s[P_i, E] = s[P_1 \& P_2 \& \ldots \& P_n, E]$,

where the circumstances are as in (4). In other words the instantial conjunction principle (4) has been derived from the equivalence and instantial comparability conditions, for the case where a single universal quantifier occurs in U^1. Nor is there any difficulty in generalising the derivation to cover cases where polyadic predicates, and therefore more than one universal quantifier, occur in U^1.

The instantial conjunction principle has strong anti-probabilistic implications, which will be discussed in §3 below. It will therefore no doubt be repugnant to many, at least at first sight. But it has now been shown that those who do not like it have to choose between repudiating the equivalence principle or repudiating the instantial comparability condition or both. For reasons already given the equivalence principle is not to be lightly impugned. So for those who object to (4) it is (3) that is the most obvious target of attack. Yet some such connections as (3) asserts between inductive support for generalisations and inductive support for their instances seems undeniable.

A possible escape-route might appear to be suggested by Carnap's definition of what he called 'instance confirmation'.[1] According to this definition the instance confirmation of a singly quantified first-order universal proposition U on the evidence E is equal to the ordinary degree of confirmation, on E, of any substitution-instance of U that refers to an individual not mentioned in E, though U's ordinary degree of confirmation, according to Carnap's theory, is normally very different from its degree of instance confirmation on E. Correspondingly it might be supposed that the plausibility of (3) rests on an equivocation whereby a single support-functor signifies both a support-function appropriate to universal propositions, and a quite different one appropriate to their substitution-instances. For if $s[U,E] = s'[P,E]$, wherever P is an instance of U, we shall certainly have $s[U,E] > s[U',E]$ if and only if $s'[P,E] > s'[P',E]$, where P' is a substitution-instance of U'. It might therefore be held that (3) can safely be rejected.

There are however two considerations that effectively block off this escape-route, because of the price that has to be paid for entitlement to use it.

First, the logical derivation of (4) from (1) and (3) is unaffected by the issue whether or not the expression '$s[H,E]$' signifies one function when H is universal and a different function when it is singular. So yet further revision of (1) or (3) would be required, if the derivation of (4) is to be prevented. For example, the scope of (3) might be restricted to singly quantified generalisations with monadic predicates, doubly quantified generalisations with dyadic predicates, and so on. Alternatively the equivalence principle (1) might be restricted in some way that will bar the substitution of U^i for U^1 where $i > 1$. But it is very difficult to see what justification could be given for either restriction other than the question-begging reason that otherwise (4) is deducible from (1) and (3).

Secondly, we can derive both (3) and (4) from (1) alone, for a large and centrally important class of propositions, provided that we may make a simple and quite plausible assumption. Let us assume that the meaning of the term 'cause' in experimental science is given by a non-contingent (i.e. relatively *a priori*) assumption as implying: same cause, same effect. Then the relation of equivalence invoked in (1) will hold between any U and any P where P is a substitution-instance of U and U is a universal proposition asserting of any event that, if it is of a certain specified kind, it causes an event of a certain other specified kind—so far as we either suppose such causal statements to specify all the causally sufficient

[1] *Logical Foundations of Probability*, p. 572.

conditions in each case, or alternatively confine ourselves to considering causal laws only in relation to circumstances that are the normal ones for applying them. But for any U and P in this situation, s[U, E] = s[P, E], according to (1). Hence for any U, U′, P and P′ in this situation, (1) implies the truth of (3). Also any two substitution-instances of U, P_i and P_j, will be logically equivalent to one another, and therefore P_i will be logically equivalent to the conjunction P_1 & P_2 & ... & P_n; so that (4) will follow directly from (1).

It looks as though a philosopher who objects to (3) and (4) in general must in any case accept them in relation to the important class of causal propositions in experimental science; and if he accepts them there he can have little motive any longer for refusing them elsewhere. If he does not want to be driven into this corner, he must find grounds for rejecting the otherwise apparently innocuous analysis of the experimentalist's concept of cause in terms of uniformity or of uniformity in normal circumstances.

Admittedly we may sometimes feel there to be more support for the hypothesis that a speck of dust in the carburettor caused our car to break down than for the hypothesis that such a speck will cause our neighbour's car, or any car, to break down. But we do so because we sometimes find it convenient to assert the former hypothesis in a sense of 'cause' that permits us to omit any mention, however imprecise or unspecific, of the other circumstances that had to be present for the speck to make trouble, and then, when we come to consider other cases, we have to reckon with the possibility that those circumstances are not present there. But this is not the experimentalist's sense of 'cause', with which the present argument is concerned. If, at least in principle, experiments are always repeatable, so too are the causal sequences that occur in them.

Hereafter, therefore, it will be assumed that (1), (3) and (4) are philosophically irrebuttable principles in the logical syntax of inductive support. Indeed, by the above-mentioned argument about causal propositions (3) and (4) are obtainable from (1) without even any restriction on E. So that, if we accept (1), we must conclude that the restriction on E in (3) and (4) was in any case superfluous for a large class of cases and perhaps in every case. The restriction on E was designed to obviate the objection that the fact described by P′ in (3) might be included in the evidence E′ for U′. But the fact that event *b* followed event *a* in an experiment gives no more support to the proposition that *a* caused *b* than it does to the proposition that another event *a*′ caused *b*′, where *a*′ is similar to *a* in all relevant respects and *b*′ to *b*. Of course, if we assume logical implication to be a special case of inductive support we shall still need to retain the restriction on E in relation to non-causal hypotheses. So further consideration of this issue must be postponed until the relationship between inductive support and logical implication is discussed in §8 (p. 61) and in §§19–22 below.

§3. Can the instantial conjunction principle be mapped on to the probability-calculus?

If an inductive function, s[H, E], satisfying the instantial conjunction principle, is to have a logical syntax that is mappable on to the mathematical calculus of probabilities, it must be a function, in effect, of p[E, H], p[H] and p[E]. But it is demonstrably not this except under one or other of certain restrictive conditions, and each of these conditions precludes its viability unless implausibly favourable assumptions are made.

The instantial conjunction principle (4) presents very obvious problems for any theory that treats s[H, E] simply as a probability of some kind, since the multiplication-law for the probability of a conjunction ensures that, if $1 > p[P_i, E] = p[P_j, E] > 0$ and $1 > p[P_i, P_j] > 0$ (which would presumably be a common case, in relation to non-causal hypotheses), then $p[P_i \& P_j, E] < p[P_i, E]$. But perhaps all that follows from this is that s[H, E] is some function of relevant probabilities (and not just a simple probability) as has often been suggested? An improbability, perhaps, or a logarithm of an improbability, or the difference between a prior and a posterior probability? It can, however, be shown that, if inductive functions are to satisfy the instantial conjunction principle, their logical syntax cannot be mapped at all on to the mathematical calculus of probabilities. The method of showing this will be first to prove that, if an inductive function satisfying (4) is to have a syntax mappable on to the probability calculus, it can have this only under one or other of certain conditions, and secondly to argue that each of these conditions is too restrictive for an inductive support-function which satisfies it to be viable.

If the syntax of an inductive function s[..., ---] is to be mapped on to the probability-calculus, any expression assigning a value to that function, for given arguments, must be translatable into some well-formed expression of the calculus which has precisely the same syntactical liaisons. But this will be possible only if for all H and all E s[H, E] takes as its value some mathematical function of one or more of the possibly relevant probabilities. At the simplest s[H, E] might take the same value as, say, p[H, E]. But p[E, H], p[H] or p[E] might also be relevant; and, since conjunction and negation are expressible in the calculus, we must reckon with the possibility that, if I is any truth-function of H or E, p[I] too might be relevant. At first sight this seems to present a very wide range of possibly relevant probabilities, some mathematical function of some or all of which may be denoted by 's[H, E]'. But in fact the list of possibly relevant probabilities is easily reduced to include just p[E, H], p[H] and p[E], when, as normally, $p[E] \neq 0$.

As a first step in the reduction it is to be noted that according to the calculus p[not-H, E] and p[not-H] are functions of p[H, E] and p[H] respectively. Also p[H & E, E] is a function of p[H, E], and p[H & E] of p[H] and p[E, H]. Again, the calculus gives us,[1] when $p[E] \neq 0$,

$$(10) \quad p[H, E] = \frac{p[E, H] \times p[H]}{p[E]}$$

So p[H, not-E] is then a function of p[E, H, p[H] and p[E]; and, of course, $p[H, H \& E] = 1$. This leaves us with just p[H, E], p[E, H], p[H] and p[E], and (10) again entitles us to drop p[H, E], when $p[E] \neq 0$.

Now let us consider the simplest non-trivial case of the consequent, of (4), viz.

$$(11) \quad s[P_i, E] = s[P_i \& P_j, E]$$

We do not need to assume here that the domain of the universal proposition of which P_i and P_j are instances is infinite, or that it is finite but unbounded, or even that it is bounded but not exhaustively catalogued. All that is required is that this domain should have at least two identifiable members, so that P_i and P_j are distinct propositions.

It will be convenient to write 'a' for '$p[E, P_i]$', 'b' for '$p[P_i]$' 'c' for 'p[E]', 'u' for '$p[E, P_i \& P_j]$', 'v' for '$p[P_j, P_i]$' and 'w' for '$p[P_j, P_i \& E]$'. Then we seek to know the conditions under which 's[H, E]' can be written as a function of three variables that, in the appropriate circumstances, satisfies (11). In other words we seek to know the conditions for the function f under which

$$(12) \quad f(a, b, c) = f(u, bv, c).$$

Moreover since the probability calculus gives us,[2] when $p[P_j, P_i] \neq 0$,

$$p[E, P_j \& P_i] = \frac{p[E, P_i] \times p[P_j, P_i \& E]}{p[P_j, P_i]}$$

we can transform (12), when $v > 0$, into

$$(13) \quad f(a, b, c) = f\left(\frac{aw}{v}, bv, c\right).$$

[1] Cf. H. Reichenbach, *The Theory of Probability* (1949), p. 91.

[2] Ibid., p. 90. In fact this is derivable from (10) with the help of the theorem for conjoint probabilities.

The first thing to be noted about any function f that satisfies (13), where a, b, c, v and w are all real numbers in the closed interval $[1,0]$— as all values of probability-functions are —, is that under certain conditions f is independent of its second argument. For under certain conditions we can prove

(14) $f(a, b_2, c) = f(a, b_1, c)$

where $1 \geqslant b_2 > b_1 > 0$. The proof rests on the assumption that if f is not independent of its second argument it is not independent of that numerical variable on any choice of values for v and w in (13), and it will be shown that it is independent of its second argument—i.e. that (14) is true—on some choices of values for v and w. Choose $v = w$, and put $v = b_1/b_2$ and $b = b_2$ in (13). Then we obtain (14) from (13) and so from (12), where $1 > v > 0$ and therefore $b_2 > b_1 > 0$. Thus, under the condition $1 > v > 0$, f is independent of its second argument.

Secondly, it is to be noted that under certain conditions f is independent of its first argument. For under certain conditions we can prove

(15) $f(a_2, b, c) = f(a_1, b, c)$

where $1 \geqslant a_2 > a_1 > 0$. Two cases will be considered: where $v > w$ and where $v < w$. Case (i), where $v > w$: put $w = va_1/a_2$ and $a = a_2$ in (13) and we obtain

(16) $f(a_2, b, a) = f(a_1, bv, c)$,

where $1 \geqslant v > w > 0$ and therefore $a_2 > a_1 > 0$. Then from (14) and (16) we obtain (15), where $1 > v > w > 0$. Case (ii), where $v < w$: put $v = wa_1/a_2$ and $a = a_2$ in (13) and we obtain (16) where $1 \geqslant w > v > 0$ and therefore $a_2 > a_1 > 0$. Then from (14) and (16) we obtain (15) where $1 \geqslant w > v > 0$. So from case (i) and case (ii) together we can infer that f is independent of its first argument under the conditions $v \neq w$ and $1 > v > 0$ and $1 > w > 0$.

It follows that f is dependent only on its third argument where $v \neq w$ and $1 > v > 0$ and $1 > w > 0$. That is, if s[H, E] is some function, f, of possibly relevant probabilities for the circumstances specified in the antecedent of (4), it turns out to be a function only of p[E] (when p[E] $\neq 0$) unless either $v = w$ or $v = 1$ or $v = 0$ or $w = 1$ or $w = 0$. But, if s[H, E] is a function of p[E] alone, s[H, E] = s[I, E] for any H, I and E, and a function which thus implies that any given piece of normal evidence gives the same support to each of any pair of propositions is blatantly worthless. Any viable inductive support-function must avoid being a function of p[E] alone. Hence any viable inductive support-function that is mappable on to the probability-calculus presupposes a probability-metric such that in the circumstances specified in the antecedent of (4) one or more of the following conditions, (17) to (21), holds good:

(17) $v = w$, i.e. $p[P_j, P_i] = p[P_j, P_i \& E]$, or
(18) $v = 1$, i.e. $p[P_j, P_i] = 1$, or
(19) $v = 0$, i.e. $p[P_j, P_i] = 0$, or
(20) $w = 1$, i.e. $p[P_j, P_i \& E] = 1$, or
(21) $w = 0$, i.e. $p[P_j, P_i \& E] = 0$.

But (19) requires that the truth of any one substitution-instance of a generalisation certifies the falsehood of any other, which is absurd. It is scarcely less absurd to require, as does (21), that the conjunction of any evidential statement with any one substitution-instance of a generalisation certifies the falsehood of any other substitution-instance of that generalisation. According to (17) no evidential statement can increase the probability of one substitution-instance on another, and this is absurd unless the latter probability is already maximal, as in (18), and thus (20) holds good. But it cannot be the case that, as in (20), any evidential statement whatever, E, certifies P_j, or even that a conjunction of E and P_i does this, unless (18) is true in any case, since we must be able to allow the possibility that E is highly adverse to P_j.

It follows that a viable inductive function that is mappable on to the probability calculus can satisfy (4) only in the context of a probability-metric that assigns a maximum value, as in (18), to the probabilities on one another of any two substitution-instances of a given generalisation, whatever other evidence there may be for or against them. But such a metric requires every generalisation to be true in all cases or true in none. It precludes us from supposing that a generalisation may be true in some cases and false in others. Yet we often want to be able to say that a generalisation held good under certain tests, and then failed in others.

There is indeed one important type of generalisation that does seem to have the property required by (18). For, as was pointed out in §2, the substitution-instances of a causal generalisation are necessarily all equivalent to one another. But there are at least three reasons why this fact does not weaken the thrust of the anti-probabilist argument developed above. The first reason is that most philosophers who have ascribed a probabilist syntax to inductive support for singular propositions have either not confined, or even—like Carnap—not allowed, the application of their theories to causal propositions. The second reason is that it is unsatisfactory to have to settle for a syntax of inductive support in the case of causal hypotheses that is radically different from the syntax relevant to non-causal ones—e.g. to hypotheses about the co-occurrence of certain kinds of biological features. Ideally one would hope to find, as a syntactic definition of inductive functions, a definition that is as unrestricted as possible in relation to subject-matter. The third reason is that the conjunction principle for universal hypotheses, which is argued in §§8–9 below, has the same pattern that obstructs any probabilist theory unless, for any two propositions H and I, p[H, I] = 1. Acceptable as this equation may be where H and I are both substitution-instances of a causal generalisation, or even where they are both instances of a generalisation of some other kind, it is obviously unacceptable where they are logically and causally independent generalisations or instances of such.

Perhaps a defender of probabilist syntax for inductive support will object that the attack mounted against it rests on the supposed intelligibility of statements about certain absolute, or antecedent, probabilities, viz. about p[H] and p[E] in (10). Notoriously, he might claim, great difficulties face any attempt to determine the values of such functions for particular arguments.

Such an objection, however, is of no avail. The outcome is no different if we confine ourselves to relative, or posterior, probabilities and think of 's[H, E]' as denoting a function of just the two probabilities p[H, E] and p[E, H]. Instead of (13) we then have

$$(22) \quad f(a, b) = f\left(\frac{aw}{v}, bw\right)$$

where $b = p[P_i, E]$ and a, v and w are as before.

But by proofs similar to those given above it can be shown that the function f is independent of its first and second arguments under just the same conditions as before. As a viable measure for degree of inductive support such a function would be no improvement at all on one that was based in part on absolute, non-relative probabilities.

Finally, it may be objected that the attack mounted here against probabilist theories of inductive syntax hits only those theories that take inductive support to be a function of one or more logical probabilities. The attack does not hit, it may be claimed, those theories that take inductive support to be a function of one or more statistical probabilities or relative frequencies, since such probabilities do not take propositions, but sets of events, as their arguments, and consequently they do not admit of the move from (11) to (12).

Such an objection cannot afford much consolation to devotees of theories, like Popper's or Carnap's, that invoke logical probabilities in one way or another—i.e. theories that make some essential use of the probability-functor and require this functor to take expressions denoting propositions, not sets, as fillers for its argument-places. Nevertheless the objection might seem to suggest grounds for concluding that what is wrong is not the wider thesis that inductive support is some function of *some* relevant probabilities but the narrower thesis that inductive support is some function of relevant *logical* probabilities. Against that suggestion only two things can be said in the present chapter.

First, there is at least one very obvious proposal for a statistical measure of inductive support that is certainly hit by the anti-probabilist attack that has been mounted here, viz. the proposal to measure E's support for the hypothesis, H, that any given member of a particular population has a certain specified characteristic, S, by the relative frequency of S in this population that is estimatable from E. For Carnap has shown[1] that in normal cases, and in a well-defined sense of 'estimate', the logical probability of H on E, as he measures it, is equal to the relative frequency of S as estimated on the basis of E. Hence so far as E's inductive support for H cannot be a function of p[H, E] it cannot be a function of the relevant relative frequency estimatable from E.

Secondly, the sheer strength and generality of (1), (3) and (4) require that, if the syntax of inductive support is to be mapped on to the probability calculus in any way at all, it must be on the assumption that probability-functions are taking whole propositions for their arguments. Otherwise (4), and therefore also (1) or (3) or both, would have to be replaced by much weaker principles, formulated in terms of propositions about relations between sets or properties, in order that the syntax of support-functions could be mapped on to that of probability-functions which were assumed to be taking sets as their arguments.

But these two points may not altogether outweigh the undoubted temptation that exists to seek some dependence of inductive support on statistical probabilities, and therefore to conceive the logical structure of inductive support-functions in a way that will admit of mapping on to the mathematical calculus of probabilities. Many probabilities are easy to measure or estimate within statable degrees of accuracy, their mathematical syntax is well understood, the importance of statistical probabilities in science is undeniable, the term 'probable' itself seems to carry just the required implication that something has at least partial but not necessarily total justification, and so on. Hence it will be necessary in a subsequent chapter (§§12–15) to say something about the role actually played in science by statistical probabilities, and by certain logical probabilities that are functions of them, and to clarify how this role differs from the role of what is here being called inductive support. Such a clarification should be taken as a necessary reinforcement of the syntactical theses advanced in the present chapter.

[1] *Logical Foundations of Probability*, p. 168 ff.

§4. The problem of a conjunction principle for universal propositions

It does not seem possible to derive an adequate conjunction principle for universal propositions from the syntactic principles already discussed. It can be shown, on certain assumptions, that the syntax of a viable inductive support-function for generalisations has no better chance of mapping on to the calculus of probabilities than has the syntax of a viable support-function for their substitution-instances. But to obtain more positive information about the syntax of inductive support for universal propositions we shall need to look at the way in which such support may be assessed.

What has been shown so far is that there are good reasons for taking the logical syntax of inductive support-functions to include the instantial conjunction principle and for supposing therefore that the mathematical calculus of probabilities cannot suffice to articulate this syntax. But even if inductive support for certain singular propositions—viz. the substitution-instances of first-order generalisations—cannot be a function of relevant probabilities, nevertheless inductive support for generalisations themselves might be such a function. It is worth while seeing whether the present line of reasoning, from purely syntactical premisses like (1) and (3), can be extended in a way that will settle this issue.

One possibility worth considering is that a conjunction principle can be established for universal propositions that in appropriate conditions produces the same probability-resistant structure as (11). But in fact it is not possible to establish such a principle along the present line of reasoning.

We can certainly prove that one generalisation has the same level of inductive support from the evidence as has its conjunction with certain others, so long as all the generalisations concerned are equivalent to substitution-instances of some single higher-order generalisation and the report of the evidence—analogously to E in (4)—does not relate specifically to one of the generalisations any more than to each of the others. For then[1] an instantial comparability principle may be assumed to apply also to higher-order generalisations, and the proof goes through exactly as for the substitution-instances of a single first-order generalisation in §2. For example, a universal proposition like 'If lions are vertebrates, they have kidneys' may be treated as logically equivalent

[1] But this condition on the evidence for higher-order generalisations does not normally hold, and an instantial comparability principle is not normally valid for higher-order generalisations: cf. §9 below.

to a proposition like 'If the lion species is a species of vertebrate, it is a species with kidneys'. Hence, 'If lions are vertebrates they have kidneys' and 'If tigers are vertebrates, they have kidneys' are both equivalents of substitution-instances of a single higher-order generalisation, viz. 'Any species of vertebrate is a species with kidneys,' and each proposition has the same level of support from appropriate evidence as has their conjunction.

Nor is the relationship that is described in (4) limited to substitution-instances of a single first-order generalisation. It applies also to those that are equivalent to substitution-instances of different first-order generalisations, provided that each of the latter has the same level of support from the evidence as its conjunction with the others. For we can show that the consequent of (4) holds good where $P_1, P_2, \ldots P_n$ are substitution-instances of generalisations each of which has the same level of support from the evidence as its conjunction with the others. E.g., if 'All ravens are black' has the same level of support from E as has 'All ravens are black and all swans are white', then it also has the same as "All x and all y are such that x is a raven only if black and y is a swan only if white', and the required conclusion is again derivable as in §2.

But these further developments are of relatively little interest. They are principles that are only applicable under very special conditions. Having therefore drawn blank in this direction we might seek to by-pass the problem of a conjunction principle for universal propositions. Perhaps some conclusions about such propositions may in any case be drawn from the thesis that inductive support for their substitution-instances is not a function of any of the relevant probabilities, if a direct correlation of some kind can be established between s[U, E] and s[P, E] where P is any substitution-instance of U.

In the case of causal propositions, as remarked in §2, such a correlation is easily obtained from (1). We have in the case of causal propositions a uniformity principle

(23) $s[U, E] = s[P, E]$,

where P is any substitution-instance of U; and, if (23) held good for non-causal propositions also, we could easily construct an argument to show that support for first-order universal propositions is not a function of relevant logical probabilities. For, if s[U, E] is some function of its relevant probabilities, (10) entitles us to suppose that it will be a function of p[E, U], p[U] and p[E]. Presumably p[E, U] is not independent of p[E, P], and the latter if not a function of p[E, U] alone, is at least a function of p[E, U] and the size of U's domain—where E, P and U are as in (4). Similarly p[P] may be presumed to be a function either just of p[U] or of p[U] and the size of U's domain. From these two presumptions it follows that so far as s[P,E] is independent of p[E,P], it is also independent of p[E,U] where E is as in (4), and so far as it is independent of p[P] it is also independent of p[U]. Hence (23) entitles us to conclude that s[U,E], where E is as in (4), is independent of all its relevant probabilities except p[E], under the same conditions as s[P,E] is independent of all its relevant probabilities except p[E]. I.e., if (23) holds good for non-causal propositions, the syntax of a viable support-function for first-order generalisations has no better chance of mapping on to the calculus of probabilities than has the syntax of a viable support-function for their substitution-instances.

But is the validity of (23) demonstrable from (1) and (3) outside the special case of causal propositions? We can obviously derive (23) from the instantial conjunction principle (4), and thus from (1) and (3), where the generalisation U has a finite domain. But generalisations in experimental science cannot always be relied upon to have a finite domain. Where U's domain is not finite, the best we seem able to do is to derive (23) from (3), on the assumption that s[H, E] has a finite and well-ordered set of values $x_1, x_2, \ldots, x_n$. On that assumption we cannot afford to ignore the logical possibility that some finite set of generalisations $U_1, U_2, \ldots, U_n$ will turn out to be such that s[U_1, E] $= x_1$, s[U_2, E] $= x_2$, . . . , s[U_n, E] $= x_n$; and the instantial comparability principle (3) is only tenable in the face of this possibility if (23) is also true. For if (23) were not true in these circumstances, there would be no room for the requisite inequalities to hold between levels of support for substitution-instances as well as between levels of support for generalisations.

But again we have arrived at a conclusion of relatively limited interest. First, our overall conclusion is negative in character, ruling out a probabilistic syntax in the case of first-order generalisations, rather than stating positively what kinds of syntactic principles do apply to these propositions. We still know nothing positive about the relation between s[U,E] and s[U & U′,E], for example, where E,U and U′ are as in (3). Secondly, the validity of (23), and with it the validity of our overall conclusion, rests on the as yet quite unargued assumption that every inductive support-function has a finite and well-ordered set of values. Thirdly, as a move against probabilistic theories our argument is almost question-begging, in so far as it rests on an assumption which implies that the values of an inductive support-function are not the members of a section of the continuum of real numbers.

We therefore seem to need some other foundation than just (1) and (3) in order to establish anything further of importance about the logical syntax of inductive support; and unless more principles can be established, or at least a principle of considerably greater generality than (4), we can hardly hope to construct a systematic theory—a calculus of inductive support. But what other premisses are available for the informal, philosophical establishment of such syntactic principles? In a field where syntax is itself at issue it is dangerous to rely on any syntactical premisses that are not at least as irrebuttable as (1) and (3). So we are forced to look instead at the semantics of the inductive support-functor.

The next chapter will pursue this alternative line of enquiry. By examining the way in which inductive support is, or can readily be, assessed in familiar types of case we shall be able not only to discern syntactic principles by which, under certain conditions, one such assessment places implicit restrictions on another, but also, incidentally, to justify the assumption that inductive support-functions should be conceived to take a finite and well-ordered set of values. However, there will then be some risk that what we are tempted to treat as a syntactic principle, valid for all inductive functions whatever their arguments—and therefore available as a defining condition on the concept of an inductive function — is better treated as a feature of inductive semantics that has no wider validity than the types of support-assessment on which our argument is actually being based. It will therefore be necessary to show, so far as our argument is now to have a semantic foundation, that the law or laws we propose for the role of syntactic principle are also implicit in other, widely different types of inductive assessment. This will be shown in Chapters III, IV and V.

II

The Ranking of Experimental Support

§5. The concept of an experimental test

To be able to test a universal hypothesis U experimentally we require to assume that certain natural variables are inductively relevant for any member of a set of materially similar hypotheses to which U belongs, where a natural variable (or variable, for short) is conceived of as a set of observationally identifiable circumstances of which the descriptions are contrary or contradictory to one another. Our assumptions about inductive relevance (or relevance, for short) are themselves empirically grounded and empirically refutable, and our criteria of material similarity are also alterable in the light of empirical considerations. A test consists in trials of a hypothesis to determine whether or not it is positively instantiated in various possible combinations of variants of relevant variables. But though the results of such tests are often said to give more, or less, support to particular hypotheses, no exact method of scoring such results is in use among experimenters. A superficially plausible proposal would be to score each favourable trial outcome as $+1$ and each unfavourable one as -1, and to measure the test's support for the hypothesis by the ratio of the summed trial-outcome scores to the total number of different trials that are possible on given assumptions about relevant variables. But this proposal encounters four major difficulties: test-results on materially dissimilar hypotheses are doubtfully commensurable, trials may differ in level of complexity, some variables may be more relevant than others, and favourable trial outcomes are bound to be assymmetrical with unfavourable ones in important respects.

The concept of induction with which this book is concerned is anchored to the discourse of experimental science. No syntactic definition of an inductive function will count as adequate unless it at least fits valid assessments of the support given to a universal hypothesis by statements reporting the outcome of experiments designed to test it. But what is an experimental test?

Let us consider at first, for the sake of clarity, only very elementary first-order descriptive generalisations, of the form 'Anything, if it is R, is S', where 'R' and 'S' describe kinds, characteristics or circumstances, instances of which may be observationally identified (with or without the aid of instruments such as microscopes, balances, clocks, thermometers, etc.).[1] The account given of these hypotheses can later be extended to cover hypotheses that are richer and more realistic in certain respects. Now, to test such a hypothesis is to investigate whether there is some R to be found that is not S. But under what circumstances should such an investigation be carried on? It is often possible to report a situation in which the hypothesis was positively instantiated, i.e. in which some R was S. However, such reports do not constitute the report of an experimental test unless the circumstances of the situation were at least to some extent deliberately selected or contrived[2] because of the inductive relevance (or relevance, for short) of certain natural variables. Not all swans are white, we know. But are all ravens black? To test whether they are we should perhaps arrange to look at ravens in summer and in winter, say, in hot countries and in cold, in youth and old age, in the mating season and the non-mating one, and so on. It would be no test of an ornithological hypothesis about plumage-colour just to go and look in the nearest park on the first afternoon that happens to be free. Similarly it is no test of a drug's non-toxicity just to pour some of it into the college cat's milk and see whether the cat survives the night. Normally we should at least[3] need to vary the size of the dose, say, and to vary the circumstances of our tests from one animal species to another and from oral to intravenous administration.

Let us term any set of observationally identifiable kinds, characteristics or circumstances of which the descriptions are either contrary or contradictory to one another, a 'natural variable', or 'variable' for short. For example, summer (i.e. being in summer) and winter constitute a natural variable, so do spring, summer, autumn and winter, and so also do summer and not-summer. Some sets of kinds, characteristics or circum-

[1] Even the most elementary hypotheses actually entertained in modern experimental science are, of course, rarely as simple as this. In particular such elementary hypotheses very often mention magnitudes that require statistical estimation, such as probabilities, means, variances, etc. But it would only serve to confuse matters if we considered such hypotheses at the outset. It is convenient to operate for a while at a certain level of analytical abstraction, and to postpone till §13 below the consideration of how statistical estimation fits in with our theory of inductive support.

[2] The term 'experiment' is perhaps more often used in a sense that restricts it to contrived situations, where the experimenter has actively interfered with nature, as distinct from merely selected situations, where the experimenter has been comparatively passive and just chosen to observe in some circumstances (the relevant ones) but not in others. However, the less restrictive sense is employed here, in order not to exclude applicability to such sciences as astronomy or natural history.

[3] In certain types of experiment controls are also needed. Some of the issues raised by this will be discussed in §9 below.

stances that are natural variables by this definition are linguistically entrenched, in the sense that there is a corresponding single word or phrase in ordinary language, like 'season' or 'method of administration'. But very many natural variables are not linguistically entrenched. Some of these are nevertheless quite familiar ones, like the rat/cat/monkey variable in pharmacological experiments. But others may be quite unfamiliar ones. For example, according to the proposed definition one variable is constituted by the set of circumstances: being a dog and being an aircraft. Admittedly such oddly constituted sets of circumstances as the latter are not ordinarily spoken of as variables. But that is because they have not yet been found to be relevant for any familiar type of hypothesis. In order not to obstruct possible discoveries of this nature we must here define 'variable' very broadly, and then go on to tackle the question: what are the criteria for telling whether a variable is a relevant one for the purpose of testing a given hypothesis?

The kinds, characteristics or circumstances constituting a natural variable may conveniently be termed its 'variants'. Consider two-variant variables first. We need to have grounds for supposing the relevance of the variables before an actual test is carried out; and, if the hypothesis to be tested has never been tested before, these grounds must lie in facts that have come to light about one or more other hypotheses that are sufficiently similar in subject-matter. These other hypotheses must not mention, explicitly or implicitly, either variant of the variable. But one or more of them must both have been falsified in some situation when one variant of the variable in question was seen to be present, and also have been positively instantiated (i.e. have had its antecedent and consequent jointly instantiated) in at least one otherwise apparently similar situation when the other variant was present. Or, if such a falsification and non-falsification have not been actually observed, then they must be predictable consequences of a scientific theory or theories which themselves have some inductive support on available evidence. For example, variation from winter to non-winter might be thought relevant for testing hypotheses about animal species' colour, if the hypothesis that all hares are always grey is believed to be normally falsified in winter when some hare-fur is white, and to hold good at other times. Or variation from a lime-free soil to one with a high proportion of lime, say, might be thought relevant for testing a hypothesis about plant-growth, such as the hypothesis that heathers are hardy perennials, because the growth habits of at least some plants (e.g. rhododendrons) are affected by this variation.

What should we say about variables containing more than two variants? It seems that a variable with n variants, where $n > 2$, is normally held to be relevant for any member of a particular class of materially similar hypotheses if and only if each variant is also a variant of one or more two-variant variables that are relevant. But clearly there must be some empirically based restriction on the worth-while range of variation if tests are not to be indefinitely extensive. In relation to each variable there must be a definite, humanly manageable number of trials in any complete test. So it will be convenient to require also that no two variants of the same relevant variable should have falsified, or positively instantiated, all and only the same hypotheses as one another. For example, one of the variables customarily described as relevant for certain hypotheses may be a physical parameter with a continuum of values, like volume or temperature. But in practice experimenters must normally have in mind a relatively small, finite set of intervals on such a continuum, within each of which instantiations of a hypothesis are to be sought. Hence, for the purposes of inductive logic, it is this set of intervals that is most conveniently spoken of as the relevant variable.

Nothing has been said so far to bar one relevant variable from being included in another, in the sense that all the variants of the former are also variants of the latter though not all variants of the latter are variants of the former. But it would be pointless for a test to manipulate such an included variable as well as the variable that includes it. So henceforth, for the purposes of inductive logic, it is safe to assume that in listing, or counting, all the relevant variables for hypotheses of a certain type we should pass over any relevant variable that is wholly included within another relevant variable. Similarly, if one relevant variable consists of two contradictory variants V^1 and not-V^1, while another consists of n contrary variants V^1, V^2, ... and V^n, the former should be ignored in favour of the latter; and also, if one variant of a relevant variable is a special case of some variant of another relevant variable, the latter variant should be ignored in favour of the former.

It is important to note, however, that we may sometimes discover that we were wrong in supposing a particular variable to be relevant for testing certain kinds of hypotheses. Here, as anywhere else in experimental science, we may have overlooked a vital factor in the situation. Perhaps we first thought we had a case of V in which the hypothesis U was falsified, and a case of not-V, otherwise apparently similar, in which U was positively instantiated. So the variable V/not-V was put down as relevant for all hypotheses materially similar to U. But then we discover a case of not-V, otherwise apparently similar, in which U is falsified, and this stimulates us to discover a feature of dissimilarity between the first two cases—a hidden variable represented by one of its variants in the first and third cases and by another in the second. Perhaps it isn't really the season of the year, but rather the temperature, or the temperature combined with the species, that makes a difference to hares' colour. Sometimes too we may acquire altogether new beliefs of this type and come to think that another variable is relevant for certain hypotheses, besides the variables previously thought so. The thalidomide tragedy has led very many to believe, who did not believe before, that the pregnancy or non-pregnancy of the experimental animal is a relevant variable for testing hypotheses about a drug's non-toxicity. Nevertheless in testing a hypothesis at any one time experimenters have to act on the beliefs they have at that time about what variables are relevant, and so any definite assessment of the extent to which their experimental results support the hypothesis is as much subject to correction as those beliefs are. Experimenters cannot at that time take into account what they do not then know, and may perhaps never know, about the indefinitely large number of other variables that may conceivably be relevant for testing members of that particular class of materially similar hypotheses.

But what are the kinds of criteria by which two or more universal propositions are to be judged materially similar to one another for inductive purposes? We need to have grounds for supposing material similarity between hypotheses that do not depend on the outcome of testing those hypotheses. Otherwise we should not be able to say what variables were relevant for a hypothesis without having tested it already, and so, since designing tests for a hypothesis requires us to presume what variables are relevant for it, no test would ever take place. It follows that material similarity must always be linguistically defined—by reference to a set of non-logical terms. By the expression 'a particular set of materially similar propositions' we can conveniently understand, for the moment, a set containing all and only those propositions in which all the non-logical terms (i.e. descriptive terms) are members of a particular set, or semantical category, of non-logical terms and the domain of discourse is a specified set of elements to which these terms may be significantly applied.

But how are such categories of non-logical terms to be determined? Names for variants of a linguistically entrenched natural variable will often belong in the same category, so that, for example, 'All swans are white' and 'All ravens are black' become materially similar to one another. But, if a variable is very extensive, its variants may not all be named in the same category. The names of two chemical compounds will perhaps both belong in a particular pharmacological category only in virtue of a previous discovery that they have importantly similar properties. Again, if the same variant of a particular variable falsifies two hypotheses, this may be taken as *prima facie* evidence that their non-logical terms should be grouped together for these purposes. But perhaps they should not be grouped together if no more such shared variables are found, or if a variable falsifies hypotheses using terms of the one kind in a much larger proportion of cases than it falsifies hypotheses using terms of the other kind. Also, of course, if every hypothesis that can be formulated to answer some question arising in the field turns out to be false, we have clearly set up too narrow a category of non-logical terms. Ornithology, for example, needs a fairly rich vocabulary of colour.

It is for scientists themselves, however, not philosophers, to determine from time to time the exact criteria of categorial identity for this purpose, each in regard to his own field of research. If they do not do so, at least implicitly, they have no rational basis for selecting some sets of circumstances in which to test a new hypothesis in their field rather than others —out of the indefinitely wide variety of circumstances that have been observed to affect hypotheses of different kinds.

It will therefore be assumed here that some such criteria, along with the appropriate list of relevant variables, are presupposed whenever a hypothesis is said to have so-or-so much support on given evidence. In other words, every normal judgement about inductive support is dependent on assumptions about the field of enquiry to which the hypothesis judged belongs.[1] The support-function invoked in such a judgement applies standards of evaluation determined in some way by the variables that are relevant to the particular set of materially similar hypotheses to which the hypothesis judged belongs. But it would be quite outside the scope of the present enterprise—which is solely to determine what is germane to the logical syntax of inductive support—to lay down any specific criteria of categorial identity and material similarity, or even any general pattern which these criteria must always exemplify. Indeed any attempt to lay down a general pattern would run the risk of placing an obstacle in the path of enquiry. For example, in pioneer hypothesis-testing, in new fields of research, closely drawn criteria of categorial identity cannot be of any use at all, since they bring no relevant variables into the net. Instead, in such fields very tenuous, novel or speculative criteria sometimes produce quite valuable results or at any rate are worth employing in a pilot-study.

Anyhow, enough has already been said about material similarity for us to be able to articulate the conditions under which a hypothesis may properly be said to pass a test. But for convenience of exposition it will

[1] I mean by this merely that the judgement is true only if the assumptions are also, not that the judgement itself, if fully spelt out, should specify the assumptions on which it depends. The latter view would prevent any such judgement from being formulated in a finite period of time, since the assumptions themselves take the form of a higher-order inductive assessment (see p. 92, n. 1 below).

help to make two further simplifying assumptions at the present stage, which can be taken up again later when the main features of support by experimental tests have been outlined. We can assume for the moment that no expressions describing variants of a relevant variable are to be, or be definable in terms of, members of the category of non-logical expressions determining that particular class of materially similar hypotheses.[1] We can also assume for the moment that in any particular set of materially similar universal hypotheses we need only concern ourselves with the sub-set of testable hypotheses, in a sense of 'testable' to be discussed later.[2] For example, let us suppose that the hypothesis 'Anything, if it is R, is S' is a member of this sub-set, and that it is being tested for the effect on it of just one variable, *v*. Then we need to make as many trials of the hypothesis as *v* has variants. The hypothesis has passed the test if it turns out that there is an R thing—no matter what—that is also S in each of these contrived or selected variants. The hypothesis has failed the test if falsified by any of the trials.

If, however, the hypothesis is being tested in relation to more than one relevant variable, the number of trials needed may be greater than the total number of the variables' variants. For we need to test for the effect of every possible combination of variants. For example, if just two variables are at issue and each has three variants, and no variant of either variable happens to cause the presence or absence of any variant of the other, then nine trials are needed.[3] Unless all possible combinations

[1] The adjustments that need to be made if this assumption is discarded are discussed in §9, p. 77 and §16, p. 146 below.

[2] In §11. On this account the concept of an irrelevant variable—a variable that is not inductively relevant to members of a particular class of materially similar hypotheses—is rather an unrealistic one. For it not only embraces variables that could conceivably be relevant, but happen not to be so, in the way that the hair-colour of a patient, say, happens to be irrelevant to the therapeutic efficacy of any antibiotic. It also embraces variables that could not conceivably be relevant to any member of a particular class of materially similar hypotheses, because the description of one or more of their variants is incompatible with the antecedent of any such hypothesis that is testable, in the way that the variable of being a dog or being an aircraft is irrelevant to any hypothesis about plant-growth like 'All heathers are hardy perennials'. A narrower and more realistic concept of an irrelevant variable may be obtained if we first define the concept of a compatible variable, for any member of a particular class of materially similar hypotheses, and then define an irrelevant variable, for such a class, as a variable that is compatible but not relevant. However, since the concept of an irrelevant variable is not required for the argument, the concept of a compatible variable has not been introduced into the text. It should also be noted that what has been given in the text is not intended as a definition, analysis or elucidation of inductive relevance, but as a set of criteria for detecting it. What we *mean* when we call a variable a relevant one for certain hypotheses is that this variable must be manipulated in any tests of them that are as thorough as possible. But to detect which variables are relevant we must look at how such hypotheses get falsified.

[3] More complicated calculations would obviously be required where the relevant variables were only partially independent of one another, as when some variants of v_1 are connected with variants of v_2, and some are not.

are tried out, we cannot be said to have tested the hypothesis as severely as possible for the effects of these variables, since the effects may be operative only in certain combinations and not in others.

How is such a test to be scored? Here one reaches the point at which the philosophy of inductive support cannot just cite widely recognised norms of experimental procedure. That kind of citation is available for elucidating the concept of a test itself, with its dependence on assumptions about what variables are relevant for particular classes of materially similar hypotheses. But in judging the results of such tests experimenters are normally content with merely comparative judgements, or with imprecisely formulated quantitative ones about, say, a 'lot' of support or a 'little'. To bring some kind of order or rationale into these judgements, and account for them systematically, the philosophy of inductive support has to propose a precise and currently practicable method of scoring tests, that goes beyond the explicit judgements of contemporary experimental science in respect of precision, but nevertheless accords with them in all other important respects.

It is tempting, at first sight, to suppose that we can construct a measure of the support given by a test-report E to a universal hypothesis U in terms of the numbers of appropriate trials—trials in variants of variables relevant for U—that E reports U to have passed or failed. For example, quantity of support might be measured by the ratio of the difference, between the number of different trials reported as being successful and the number reported as unsuccessful, to the total number of possible trials that are appropriate for hypotheses of U's kind.

Such a measure would certainly be both precise and practicable. It would apparently constitute a very widely applicable support-function, taking rational numbers between 0 and 1 as values. Unfortunately, however, it is only plausible on an untenable assumption of homogeneity. It presupposes that all outcomes of appropriate trials are homogeneous in value, except in so far as passing a trial contributes a positive unit of value and failing it constitutes a negative one. But there are four important ways in which this assumption is unjustifiable, and by examining them we shall see the main difficulties that obstruct any proposed *measure* for inductive support.

First, there is no reason to suppose that test-results on materially dissimilar hypotheses are commensurable. The commonest and most useful comparisons made in practice are certainly within a single field of experimental enquiry. One hypothesis that answers a particular question, or type of question, that has arisen is often said to have more support than another, on evidence that embraces tests on both; or the same hypothesis is said to have more support on the evidence of one test than on that of another. But the level of support for a hypothesis in descriptive ornithology, for example, is seldom if ever compared with that for a hypothesis in mammalian pharmacology. Only a very rare and special kind of purpose would be served by such a comparison. Correspondingly the concept of a test articulated here is one that relates a test to a class of empirically discoverable variables that are relevant for any one of a class of materially similar hypotheses. This concept seems to allow no empirical basis for cross-field comparisons between materially dissimilar hypotheses. On what grounds would it legitimately be assumed that the interconnections of factors and characteristics prevailing in one field were precisely paralleled in another—or sufficiently parallel to be similarly scored—which were not also grounds for taking hypotheses in these two fields to be materially similar? The only way to deal with this difficulty, at a stage of scientific enquiry at which some useful hypotheses still seem materially dissimilar to others, would be to suppose a plurality of inductive support-functions instead of just one. Each function would be identified by the class of hypotheses, closed under the relation of material similarity, with which it was primarily concerned. In relation to other hypotheses, as its arguments, it would always take the value zero.

Secondly, even within the same field of enquiry the outcome of one trial seems to be of greater significance than that of another, *ceteris paribus*, if the one trial is more complex than the other. A trial will be more complex if it forms part of a more thorough test, i.e. a test in which a greater number of relevant variables are manipulated. The report of the trial may then tell us not just that the hypothesis was instantiated in circumstance V_i, say, but that it was instantiated in circumstance V_i & V_j, where V_i and V_j are variants of different variables. We are then being told not just that V_i did not suffice to falsify the hypothesis, but that neither V_i nor V_j nor their combination sufficed to do this. However, if there were no other difficulties to surmount, this particular heterogeneity could also be accommodated quite simply. The value of each trial-outcome could be weighted by the number of mutually independent variables manipulated in the test of which the trial in question forms part.

Thirdly, and more seriously, one test may be capable of giving greater support than another even when the number of variables manipulated is the same or smaller, because one of these variables is especially relevant—especially successful at falsifying hypotheses of the type in question. For example, mating/not-mating is perhaps a more relevant variable than temperature in relation to hypotheses about bird-plumage; and pregnant/not-pregnant is perhaps a less relevant variable than medical history in relation to hypotheses about drug-toxicity. Hence, if E_1

reports, in effect, that U_1 has passed test t_1, and E_2 reports that U_2 has passed t_2, then any satisfactory assignment of values to $s[U_1, E_1]$ and $s[U_2, E_2]$ must pay due regard to the comparative relevance of the variables in t_1 and t_2. In other words the degree of relevance of the variables, variants of which are mentioned, constitutes an important heterogeneity between the reports of some trial-outcomes and others. Possibly there are some fields of research in which in practice this factor does not operate. But it certainly bars a general programme for measuring inductive support in a way that puts the values of all positive trial-outcomes on a par with one another, even within the same field of enquiry and where the trials are equally complex.

Fourthly, it is altogether too paradoxical to suppose, in relation to any individual trial of a universal hypothesis, that a successful outcome has just the same value positively conceived as an unsuccessful one has negatively. Consider a case, for example, where E_1 reports, in effect, that U_1 has failed one trial and passed two, while E_2 reports just that U_2 has passed one trial. According to the type of measure under consideration U_1 would be said to enjoy the same support if E_1 is true as U_2 does if E_2 is true, even though E_1 implies the falsity of U_1 and E_2 does not imply the falsity of U_2. Some philosophers may be able to persuade themselves to swallow this type of thing as a price to pay for other supposed merits in their theories. But any method of scoring tests on universal hypotheses of the form 'Anything, if it is R, is S' is bound to appear paradoxical to many if it does not somehow reflect the familiar asymmetry between favourable instances—the cases that are both R and S—which do not verify, however many of them are observed, and unfavourable instances —the cases that are R and not S—even one of which serves to falsify.

There are thus four important heterogeneities to be found among the trial-outcomes which the support-function under consideration treats as being homogeneous. This excessively *simpliste* support-function must therefore be rejected. Admittedly the first two difficulties can be fairly easily overcome if the proposed method of scoring trial-outcomes is modified along the lines suggested in each case. But the other two heterogeneities are more serious difficulties for such a proposal, and for any attempt to measure inductive support; and they force much more drastic alterations of viewpoint. The immediately following section, §6, will be devoted to showing this.

§6. Some obstacles to the measurement of support from experimental evidence

A variable's degree of a relevance for testing hypotheses of a particular kind cannot be appropriately assessed by a measure that is based on linguistic structure or on the distribution of attributes in an *a priori* model. Nor are the empirically given data on which we might seek to base a measure of relevance at all comparable with those on which measures of physical quantities, like volume or temperature, may be based. Nor is it possible to construct a probability-measure of relevance that is based on the frequency with which hypotheses have been adversely affected in the past by variables or tests of the kind in question. Also, though we need to allow the possibility of having $s[U, E] > 0$ even where E contradicts U, yet any suitable distribution of significance between successful and unsuccessful trial-outcomes seems relatively arbitrary. These difficulties cannot be met by proposing to adopt the measures that accord best with experimenters' intuitive comparisons of inductive support, since such a proposal misconceives the nature of the difficulties.

Any simple measure of inductive support that just counts positive and negative trial-outcomes is blocked, as was pointed out in §5, by the fact that one variable may be more relevant than another and therefore one trial-outcome, even if otherwise comparable, may be more significant than another. To deal with this difficulty it is natural to conjecture that the score for a trial-outcome should be weighted in accordance with the relevance of the variables of which variants are mentioned in any report of the trial. But how is such relevance to be measured?

It might be suggested that a variable's degree of relevance should be conceived to depend in some way on the linguistic standing of the predicates describing its variants. For example, in an artificial language-system the relevance of a variable v might be made to depend in some way on the number of logically independent descriptions of the universe of discourse that are incompatible with ascribing a variant of v to an individual element of the universe. Degree of relevance would then depend on such factors in the structure of the language-system as the number of different families of mutually contrary atomic predicates (e.g. the family of colour predicates or the family of zoological species-names), or the number of predicates within each family, or the number of possible modes of statement composition (e.g. truth-functional with 'not', 'either ... or', etc., or temporal with 'before', 'after', 'while', etc.).

But, though there is plenty of room for the exercise of mathematical ingenuity in devising postulates that will come to associate varying degrees of relevance with many of the atomic and molecular predicates of an artificial language-system, this form of ingenuity cannot suffice to solve the philosophical problem here. The trouble is not that this approach is apparently compatible with the same variable's having a different degree of relevance in different languages. For the demand that inductive relevance should be language-invariant can easily—too easily—be met. The postulates just need to be contrived to determine the structure of a model universe, rather than the structure of a language, and the relevance of a variable is then measured by the number of possible states of the universe that are excluded by the instantiation of one of its variants. The real trouble is that it has also to be shown how the choice between one such set of postulates and another, whether they be linguistic or model-theoretic, can be governed by experimentally observable facts. Here, as often elsewhere,[1] the distinction between *a priori* and empirical sources of justification is of much greater philosophical importance than the distinction between analytic and synthetic truth. For, when scientists working in a particular field of research judge one variable more relevant than another for hypotheses in their field, they do so on the basis of past experience in that field, and their judgement is itself open to correction by future experience. Judgements about the relative importance of trying out new drugs on pregnant animals as well as non-pregnant ones, for example, had to be revised after experience with thalidomide. The crux of the problem here is epistemological rather than mathematical, and the real issues at stake tend to be concealed, or grotesquely underemphasised, if what is offered for their solution is just an artificial language-system, or model, without any indication of how its associated measures may themselves be put to the test of experience. Such theories are like children's toys. They may create their own atmosphere of quarrel and discussion. But they do not come to grips with the real difficulty in explicating the differing degrees of relevance of different variables in a particular field of experimental enquiry.

Nor is it of any use to argue in favour of a particular artificial language-system that it approximates ordinary English, say, or ordinary Finnish in certain respects—for example, its structure of atomic and molecular predicates—since existing language-patterns are not sacrosanct in natural science; and they are not sacrosanct for two reasons.

First, it is by no means evident that the structure of a natural language at any one time either reflects or is reflected by current opinions about the differing relevance of different variables. After all, the same test may be revealed by experience to be appropriate to hypotheses about one subject-matter and inappropriate to hypotheses about another: e.g. seasonal variation in relation to hypotheses about animal colour and about drug toxicity, respectively. Yet the same natural language is used overall. Also, within the same field of research two linguistically

[1] Cf. L. Jonathan Cohen, *The Diversity of Meaning*, 2nd ed. (1966), p. 173 ff., and also §19, p. 189 below. On the demand for language-invariance, and Carnap's model-theoretic way of meeting the demand, cf. Wesley C. Salmon, 'Carnap's Inductive Logic', *Jour. of Phil.* lxiv (1967), p. 733 f., and the references there cited.

comparable families of predicates may describe variables that differ substantially in degree of relevance: e.g. the male-female and day-night distinctions in relation to hypotheses about plumage-colour.

Secondly, even if it be argued that natural language is always in course of adaptation to the needs of natural science, the same epistemological problem has still be to solved. Admittedly old parameters or families of predicates in natural language are sometimes altered, or new ones adopted, as when Fahrenheit's scale of temperature came into everyday use. But just so far as isomorphism with natural language is held a criterion of adequacy for an artificial language-system, the same difficulties arise when any attempt is made to consider the nature of the reasons why experimental scientists should, or should not, be content with the present structure of their natural language.

Of course, all that has been argued here is that no single artificial language-system, or model, can suffice to explicate the concept of inductive relevance. It is certainly conceivable that a hierarchy of such theories might contrive to deal more adequately with the epistemological issues involved. A suitable higher-order system might provide a relatively *a priori* mode of assessing the degree of support given by different evidential reports to various lower-order systems, within which in turn the support available for ordinary scientific hypotheses could be assessed in a way that did justice to the differing degrees of relevance of different variables in this or that field of enquiry.

But no such higher-order system, providing an adequate and practicable basis for assessing proposed measures of relevance, has yet been published. Indeed, certain very serious difficulties stand in the way of contriving one, or of contriving even an informal measure of inductive support that relies at some point on a practicable measure of inductive relevance.

In the first place, the construction of an adequate measure for relevance in a given field encounters a difficulty that does not face the construction of measures for standard physical magnitudes (volume, velocity, temperature, etc). Specifically, it makes rather strong demands on the activity of experimenters within the field in question. A very great deal of hypothesis-testing would need to go on in a particular field in order to establish both an appropriate standard unit or interval of relevance for that field and also a reasonably reliable method of calibrating a scale of relevance by reference to that standard—not to mention the further operations required in order to carry out the actual measurements for each variable. It is as if we had to devise a way of measuring the capacity of a set of heavy objects to break bridges of a certain type, without being able to weigh these objects independently or do anything but just try them out on as many non-trivially different bridges as we can, while the supply of bridges of the type in question was controlled by our own ability to think up non-trivially novel designs for them. In these circumstances it would hardly be surprising if no plausible and practicable measure of bridge-breaking capacity were forthcoming, just as no plausible and practicable measure of inductive relevance has been proposed in any field.

Someone may object:—'You are making the whole thing appear far more difficult than it actually is. From the sample offered by past experience we can estimate, for any given variable v, the probability that a hypothesis of a certain kind will be falsified by one or other of v's variants. This probability measures the relevance of the variable v for hypotheses of that kind.'

But though it proposes the only type of relevance-measure that seems at all practicable, this objection cannot be sustained. It assumes that the population of hypothesis-trials is homogeneous in all relevant aspects. But we need a measure of relevance only where more than one variable is relevant, and where more than one variable is relevant the population of hypothesis-trials is far from homogeneous. Trial-outcome probabilities may be heavily biased by the presence of variants of other variables in most tests.

It is therefore tempting to suggest that we should confine ourselves to measuring the relevance—call it rather severity—of a test as a whole, because then the problem of bias can apparently be overcome by estimating the probability that manipulation of this or that combination of variables will falsify a hypothesis of the type in question. We should have to give up the possibility of measuring the degree of support given by parts of tests, which the proposal to score individual trial-outcomes allowed us. But at least the degree of support afforded by a successful test-result, it might be suggested, could be equated with the improbability of such a result, as estimated in the light of previous experience with just that test for materially similar hypotheses.

Unfortunately, however, even the population of test-results is insufficiently homogeneous. The trouble is that most experimenters in this or that field may well be expected to learn from experience and propose hypotheses, as time goes on, that are more likely to survive testing in relation to notoriously troublesome variables than the hypotheses they proposed and tested earlier, or to propose hypotheses in one sub-area of the general field that are more likely to survive testing than the ones they propose in another sub-area. So the sample from which test-result probabilities was estimated could have quite a large bias. But the rate at which experimenters improve their hypothesis-proposals in a particular field will depend on too many relatively accidental factors (intelligence, age, social policy, etc.) to be regularly predictable. So the bias could not be allowed for in our calculations. No doubt the selection of hypotheses for testing could be deliberately randomised in some way. We might therefore think of basing our measure on the assumption that such randomisation had taken place in the ascertainment of relevant variables. But a measure so based would be invalid as a mode of assessing support for hypotheses that came to be advanced in the ordinary course of scientific research, where hypotheses are not randomly selected. Again, the number of materially similar (and testable) hypotheses could conceivably be made so small that in practice every hypothesis could be tested before any assessments needed to be made. But the ensuing measure would depend for its reliability on the relatively arbitrary decision to confine the number of such hypotheses in this way.

What we have been looking for is a measure that will assign cardinal values to a support-function of the form s[U, E], whatever trial-outcomes or test-results are reported by E. The quest has been baulked by the fact that a function of this kind must invoke some appropriate measure of the relevance of a variable, or of the severity of a test, whereas no such measure is available. Moreover the apparent futility of the quest is reinforced by the asymmetry of successful and unsuccessful trial-outcomes that was mentioned at the end of the previous section, §5. That asymmetry gives trouble just because there is often said to exist some support for a given universal hypothesis, even if it is known to have been falsified. It looks as though s[U, E] can have a value greater than zero even when E contradicts U, and the difficulty is to find a way of determining this value appropriately. For example, s[U, E] cannot then have the same logical syntax as a probability-function p[U, E], since where E logically implies not-U we have $p[U, E] = 0$. Indeed it cannot then plausibly be any function at all of the logical probabilities concerned, as may easily be shown.

It was argued at the outset of §3 that if s[H, E] denotes any function of the logical probabilities concerned it denotes a function of just p[E, H], p[H] and p[E], when $p[E] \neq 0$. But if we have $p[E, H] = 0$, because part of E's content is inconsistent with H, then any variation in the value of s[H, E], where s[H, E] is some function of the probabilities concerned, must be due to variations in the value of p[H] and p[E]. So the level of positive support that E gives a universal hypothesis U in these circumstances does not depend on there being any particular relation between U and that part of the content of E which does not consist in the falsificatory evidence. The hypothesis U could be about an entirely different subject-matter from that part of E, but the greater-than-zero value of s[U, E] would still be a function of p[U] and p[E]. And it is quite out of keeping with the discourse of experimental science that the kinds of tests E reports—the natural variables E mentions—should have nothing whatever to do with the extent of E's positive support for U. So accepting $s[U, E] > 0$ in some cases where E contradicts U is incompatible with treating the value of s[U, E] as a function of the logical probabilities concerned, and thus compounds the difficulty of constructing a measure of inductive relevance.

It is possible to side-step this difficulty, as Carnap in effect does in his theory of instance-confirmation,[1] by tying s[U, E] not to p[U, E] but to p[P, E] where P is any substitution-instance of U. But the trouble then is that successful and unsuccessful trial-outcomes tend to be attributed equal, though opposite, significance, and any attempt to assign a precise difference to their respective levels of significance seems relatively arbitrary.

On the other hand, while Carnap's theory treats successful and unsuccessful test-results too much alike, Popper's theory of corroboration[2] makes the difference between them too great. On his view, we cannot have $s[U, E] > 0$ at all where E contradicts U. But it seems very implausible to suppose that we must have $s[U_1, E_1] = s[U_2, E_2]$ where E_1 implies U_1 to be unaffected by nine relevant variables and adversely affected by one and E_2 implies U_2 to be adversely affected by all ten. After all, the reformulation (e.g. insertion of additional conditions into the antecedent) required in such circumstances to produce a hypothesis that was unaffected by all ten variables would normally be much slighter in the case of U_1 than in the case of U_2. In that sense E_1 shows U_1 to be nearer the truth, so far as it is known, than E_2 shows U_2 to be.

Finally, it may be objected that the proper way to deal with all difficulties of this nature—whether they are generated by the problem of relevance or by the asymmetry of successful and unsuccessful trial-outcomes—is simply to adopt the measures according best with the intuitive comparisons that experimental scientists actually make, when they compare the different degrees of support afforded by different test-results. But such a proposal is altogether too facile. It quite misconceives the nature of the difficulties here. Those difficulties are felt to be important just because we want to show how comparative judgements of evidential support—i.e. assertions of the form $s[U_1, E_1] > s[U_2, E_2]$—can be justified. If comparative judgements of evidential support in a particular field rely at some point on a measure of relevance, for example, we are asking why experimental scientists and others who make such

[1] *Logical Foundations of Probability*, p. 572.
[2] *The Logic of Scientific Discovery*, p. 268.

judgements should operate with one measure of relevance there rather than another. It is no answer to this question to say just that they do so operate. That answer is not only philosophically shallow and unilluminating. It also fails to provide any basis for deciding between conflicting intuitive judgements, in an area where intuitions are notoriously liable to conflict with one another; and it gives no guidance for the construction of relevance-measures in relatively new and undeveloped fields of experimental enquiry. Moreover, at the desired level of finesse a precise measure of inductive support ought to be able to produce some consequences that are beyond the discriminatory capacity of untutored intuition. It is a poor and rather pointless measure, whether of a physical quantity in ordinary science or of inductive support, that does not improve at all on our pre-metric discriminatory capacities. Hence, even if both of two proposed measures accord with existing judgements as far as those go, we may still need independent grounds for assessing the respective merits of the two proposals, since the ways in which they purport to improve on untutored intuition may conflict with one another.

Perhaps, despite all that has been said in the present section, someone will one day produce a practicable method of measuring inductive support that is based on an appropriate measure of relevance and an appropriate distribution of significance between successful and unsuccessful trial-outcomes. It would be rash to claim that such an event is quite inconceivable: some very surprising things have happened in the history of logic. But there are certainly very good grounds for concluding that in fact no method of this kind will ever be available, and any current approach to the problem of experimental enquiry and inductive support must be fashioned within the limits imposed by that well-grounded conclusion.

§7. The grading of support from experimental evidence

Because of the difficulties about measuring it, degree of relevance cannot be taken into account by support-functions. Hence a prudent experimenter in a particular field will use some set of tests that permit comparisons between test-results on the basis of nothing other than cumulatively increasing complexity of tests. This cumulatively increasing complexity must be determined by some order of priority among the relevant variables; and the latter order will depend ideally on a prudent experimenter's tentative conclusions, from past experience of their falsificatory achievements, about which variable is more relevant than which. Thus a support-function can be readily constructed, for any particular field, that is adequate to the task of ranking or grading, though not of measuring, the support given to a hypothesis by the results of any prudent experimenter's tests on it.

It was argued in the previous section that no measure of inductive support in any field is available that is based on an appropriate measure of relevance and an appropriate apportionment of significance to successful and unsuccessful trial-outcomes. But for what purpose would such a measure of inductive support be necessary? It would certainly be needed if we had to be able to assign positive values to s[U, E] in a very wide variety of cases, viz. if E may report the outcome of any appropriate trial or trials of U or if E may at least report the results of complete tests of U in relation to any one or more of the relevant variables. For then each of these possible outcomes or results must be suitably scored in relation to every other. Such a support-measure would therefore be needed also in order to provide a justification for comparative judgements of the form $s[U_1, E_1] > s[U_2, E_2]$ wherever E_i may be thus indiscriminate, as it were, in scraping together supporting evidence for U_i.

However, if the requisite measures of inductive support are not available, these purposes clearly cannot be achieved in any field, however desirable they may otherwise appear. Prudent experimenters may accordingly be expected in each field to adopt some selection of test-procedures that will ensure the possibility of drawing justified comparisons between their results. What would be the point in their conducting combinations of trials of which the results cannot be compared with one another because of incommensurabilities in respect of relevance and the asymmetry between successful and unsuccessful outcomes?

Hence, if we are content that our theory of inductive support should suffice to serve the purposes of prudent experimenters and of anyone who wishes to appraise the actual or possible results of such experimenters, we do not need to be dismayed by the difficulties discussed in the previous section. The difficulties are overwhelming only for those whose ambitions are excessive—for those who seek to be able to assess the results even of foolish and unprofitable methods of experimentation. Just as we must in any case give up the hope of achieving commensurability between test-results on materially dissimilar hypotheses, so too we must give up the hope of achieving commensurability between all possible combinations of trial-outcomes for materially similar hypotheses. But in neither case, it will turn out, do we give up anything that an ideal experimenter needs for the assessment of progress in his own field—i.e. for assessing which hypotheses are better supported by the same evidence or for judging that the same hypothesis is better supported by one evidential report than by another.

If the selection of evidence which our theory had to deal with was entirely outside anyone's control, we could not so lightly dismiss the difficulties in question. If the typical evidential data to be appraised in relation to a hypothesis about the plumage-colour of ravens were, say, whatever an observer had time to notice as he walked through the nearest park on a spare afternoon, we should rightly be disappointed if we could not see how to supply positive values to our support-functor for a correspondingly extensive range of fillers of its second argument-place. But the concept of induction with which this book is concerned is anchored to the discourse of experimental science, and it is in any case an essential feature of experimental science that the range of observations to be made in testing a hypothesis should be selectively guided, in the light of beliefs about relevant variables, by the aim to discover as much as possible about available support for the hypothesis. All that is being assumed here is that such an aim implies not the discovery of exactly how much support—in some quantitative sense—is available for the hypothesis, but rather the discovery of exactly how the available support compares with that available for other hypotheses in the same field or with support for the same hypothesis on evidence available at other times or places. Comparability of test-results is to be treated as one of the objectives aimed at by the observational selectivity that is integral to experimental science.

How then would a prudent experimenter proceed? One thing he must achieve is so to contrive his tests that he can compare their results not only without taking into account degrees of relevance but also without assuming that all the variables manipulated in the tests have the same degree of relevance. The only way to achieve this is to ensure that so far as one complete test differs at all from another, in a particular field, it differs only in thoroughness, i.e. only in the number of relevant variables manipulated. It follows that the list of relevant variables must be conceived to constitute a well-ordered set of some kind, and the only tests employed—the canonical tests—must be those that exploit, progressively, more and more of these variables. Test t_1 will consist of a single trial, when no relevant variable at all is being manipulated; test t_2 will consist of trials in all the variants of variable v_1; t_3 will consist of trials in all possible combinations of the variants of v_1 and the variants of v_2; ... and t_n will consist of trials in all possible combinations of the variants of v_1, the variants of v_2, ... and the variants of v_{n-1}. It follows that a hypothesis which fails t_i will normally also fail t_j, where $j > i$, and a hypothesis that is capable of passing t_j will normally also be capable of passing t_i. If a particular hypothesis did not behave like this, an experimenter would normally infer that some hidden variable was wreaking havoc with the situation and should consequently be added to his list of relevant variables. In other words, on the assumption that $v_1, v_2, \ldots, v_{n-1}$ are all the relevant variables it is unnecessary to perform t_i if a hypothesis has passed t_j, where $n \geqslant j > i$, and also unnecessary to perform t_j if the hypothesis has failed t_i. Consequently, if one hypothesis has passed t_j, and another hypothesis has passed no higher test than t_i, where $j > i$, the former hypothesis is better supported than the latter.

Translating this on to a level of schematic abstraction at which inductive logic can conveniently operate we shall treat the members of certain sets of ordinal numbers—not cardinals—as the values of support-functions for appropriate arguments. Whatever the hierarchies of tests to which our functions are sensitive, they will rank or grade support rather than measure it. Now, let us suppose 's[..., ---]' in 's[H, E]', whatever H or E may be, to express the function that assesses support for H and for any proposition materially similar to H. Let us assume that all and only the natural variables $v_1, v_2, \ldots, v_{n-1}$, in that order, are relevant to these hypotheses. Then 's[U, E] $= 1/n$' may be taken as an abbreviation for 'On E's evidence U attains to just the first of the n possible grades of positive support for U and propositions materially similar to U'. So a (dyadic) support-function maps ordered pairs of propositions on to the first $n+1$ integers that are greater than or equal to zero, where $n-1$ is the total number of relevant variables.

Expanding this in terms of our account of the hierarchy of tests $t_1, t_2, \ldots, t_n$, we shall have

(1) $\quad \mathrm{s}[(x)(\mathrm{R}x \rightarrow \mathrm{S}x), (\exists x)(\mathrm{R}x \;\&\; \mathrm{S}x \;\&\; \mathrm{T}_1 x)] = 1/n$

where insertion of '$\mathrm{T}_1 x$', meaning 'x is a subject of test t_1', implies that the individual so described is unaffected by any variant of a relevant variable. This implication is needed because if such a test of U has any point at all it is to discover whether U is falsified from within, as it were—i.e. whether being R is itself a bar to being S. But since experimental tests are normally designed to discover the effects of relevant variables, and no relevant variables are manipulated in t_1, survival of t_1 does not justify assigning anything higher than the lowest positive grade of support to a normal hypothesis.[1]

[1] But see §16, p. 152 ff., below for a certain type of special case. It is perhaps also worth noting, in order to forestall possible misunderstandings, that the criteria of inductive support for a particular type of hypothesis are defined here in terms of an independently assessed form of relevance, viz. the relevance of certain variables, whereas in some philosophical theories a concept of the relevance of given facts is defined in terms of their impact on the pre-existing level of support (e.g. J. M. Keynes, *A Treatise on*

Next, if variable v_1 has just two variants V_1^1 and V_1^2, we shall have

(2) $$s[(x)(Rx \rightarrow Sx), (\exists x)(Rx \,\&\, Sx \,\&\, T_2 x \,\&\, V_1^1 x) \,\&\, (\exists x)(Rx \,\&\, Sx \,\&\, T_2 x \,\&\, V_1^2 x)] = 2/n$$

where insertion of 'T_2x', meaning 'x is a subject of test t_2', implies that the individual so described is unaffected by any variant of a relevant variable other than v_1. This implication is needed because an experimental test of U in relation to v_1 seeks to discover whether U, if not falsified from within, is falsified *by*, not just *in*, a variant of v_1. For example, in testing a hypothesis about the air-resistance of an aircraft shape, in relation to varying engine speeds, engineers might perhaps seek to ensure that each experiment is unaffected by wind.

No doubt it is in practice often impossible or very difficult to be sure that extraneous factors have been adequately screened off, especially where no relevant variables at all are being manipulated, or only very few. Even if you are carefully varying the method of administering a drug, you still have to test it on *some* species of animal, and which species you choose may in some way affect the result you get. Even if you carefully vary the soil in which you grow a type of plant, the plant still has to grow in *some* temperature, and which temperature you choose may in some way affect the plant's growth. But this just serves to illustrate the truism that is underwritten by the insertion of 'T_2x', say, in (2): viz. the more thorough, or complex, a test is, in terms of the number of relevant variables manipulated, the more reliably interpretable are its results—i.e., the more sure we are entitled to be, even without further corroboration, that assumptions of the form 'T_ix' are true.[1]

The insertion of 'T_2x' in an evidential report like the one mentioned in (2), may also be regarded as guaranteeing the repeatability, in principle,

Probability, 1957, p. 54 ff. and R. Carnap, *Logical Foundations of Probability*, p. 346 ff.). Roughly, on the present account the criterion of relevance is prior to the criterion of support, whereas on the other account the criterion of support is prior to the criterion of relevance. The reason for this is that in the present account, as often in natural science, the relevance concerned (which I call 'inductive relevance') is relevance, for any E and any H, to the assessment of how much E supports H, whereas in the other account the relevance concerned must, I suppose, be, for any E and any H, relevance to the assessment of how much E affects support for H. However, the latter, which should go under a different name (say, 'evidential relevance'), can also be represented on the present analysis: cf., e.g., metatheorems 480 and 481 in §22 below. Metatheorem 499 shows how even canonical test-results can be evidentially irrelevant.

We shall see later (§9, p. 80) that when non-elementary generalisations come to be taken into account the number of relevant variables becomes equal to (not just one less than) the number of canonical tests and of grades of positive support.

[1] Note, however, that in the case of statistical hypotheses the truth of occurrences of 'T_ix' in an evidential proposition may be underwritten by a relatively random selection of the evidential samples (cf. §13, p. 121, below).

of the reported test-result, so far as we conceive of causes in the sense in which it is true to say: 'same cause, same effect' (cf. §2 above). For if, in a certain performance of test t_2 on the hypothesis U, the result was not affected by any factors other than variants of v_1 and circumstances mentioned in U, it must have been produced by some or all of those factors. So whenever U undergoes test t_2 the result will always be the same.[1]

Similarly, if variable v_1 has just two variants, V_1^1 and V_1^2, and variable v_2 has just two variants, V_2^1 and V_2^2, we shall have

$$\text{(3)}\quad s[(x)(Rx \rightarrow Sx), (\exists x)(Rx \;\&\; Sx \;\&\; T_3 x \;\&\; V_1^1 x \;\&\; V_2^1 x) \;\& \\ (\exists x)(Rx \;\&\; Sx \;\&\; T_3 x \;\&\; V_1^1 x \;\&\; V_2^2 x) \;\& \\ (\exists x)(Rx \;\&\; Sx \;\&\; T_3 x \;\&\; V_1^2 x \;\&\; V_2^1 x) \;\& \\ (\exists x)(Rx \;\&\; Sx \;\&\; T_3 x \;\&\; V_1^2 x \;\&\; V_2^2 x)] = 3/n$$

where '$T_3 x$' means 'x is a subject of test t_3' and implies that the individual so described is unaffected by any variant of a relevant variable other than v_1 and v_2. And so on, up to t_n. But in the case of t_n we shall not require each report of a trial to include the statement '$T_n x$'. It is not part of the *evidence* in t_n that something is unaffected by any variant of a relevant variable other than $v_1, v_2, \ldots, v_{n-1}$, since if no other relevant variables are known no steps can be taken to screen off their effects. Rather, the support-function itself has been constructed on the assumption that nothing can be affected by any variant of a relevant variable other than a variant of $v_1, v_2, \ldots, v_{n-1}$. A successful result from test t_n can therefore be said to afford full support for a hypothesis. So far as test-results are repeatable, such a hypothesis must be true. But, of course, it may always turn out, as we learn more about what are the relevant variables, that we were mistaken in our views about how to constitute the canonical tests for hypotheses of this kind, and it may then turn out also that we were mistaken in the grade of support we ascribed to certain of these. An experimenter's assessment of the grade of support that certain evidence affords his hypothesis is just as much open in principle to correction by future experience as is the hypothesis itself.

We shall also have such equations as

$$\text{(4)}\quad s[(x)(Rx \rightarrow Sx), (\exists x)(Rx \;\&\; Sx \;\&\; T_2 x \;\&\; V_1^1 x) \;\& \\ (\exists x)(Rx \;\&\; Sx \;\&\; V_1^2 x)] = 0/n$$

[1] So far as the result of a canonical test is always, in principle, repeatable, two performances of the same test on a given hypothesis can provide no better evidence for it than one: cf. metatheorem 499 in §22 below. Mere multiplicity of evidential instances does nothing to increase inductive support, though sample-size has a role in the estimation of statistical magnitudes, which is discussed in §12–13 below. For the sense in which test-results on statistical generalisations are repeatable, cf. p. 123, n. 1.

and

(5) $s[(x)(Rx \rightarrow Sx), (\exists x)(Rx \,\&\, Sx \,\&\, T_3 x \,\&\, V_1^1 x) \,\&$
$(\exists x)(Rx \,\&\, Sx \,\&\, T_3 x)] = 0/n$

For, in the spirit of scientific prudence, we shall not suppose any positive support to be derivable from the results of a non-canonical test—i.e. as in (4), where, so far as we are told, one of the individuals described is not assumed to have been screened off from extraneous factors, or as in (5), where not all the variables appropriate to the test are reported to have been systematically manipulated, with one variant of each (so far as they are combinable) occurring in each of the reported trials.

Patently the truth of (1), (2), (3), (4) and (5) is quite unaffected by whatever differences in degree of relevance may in fact exist between v_1 and v_2. But it escapes being affected by degree of relevance because we have been content to systematise only that concept of inductive support which is attributable to a prudent experimenter, and a prudent experimenter will conduct his tests by manipulating cumulatively increasing combinations of relevant variables, in order to ensure commensurability.

How then is the order of the set of relevant variables to be determined? What should be the well-ordering relation? Strictly so far as concerns the possibility of making some systematic comparisons of evidential support, it does not matter what well-ordering relation is chosen. The variables could be ordered by the differing monetary costs of controlling them in single-variable tests, for example, or by alphabetical priority of the initial letters in their shortest English descriptions. Whatever the basis on which the set of relevant variables is ordered, a set of canonical tests can be constructed accordingly, and the appropriate ordinal number > 0 can be assigned as the value of our support-functor for each report of successful test-results on a particular hypothesis. But, if the variables are ordered differently from the way in which a prudent experimenter would order them, the list of canonical tests will not include all and only those tests that a prudent experimenter would administer. Correspondingly, the theory of evidential support that has been constructed will not suffice to determine the truth-value of all the comparative judgements about evidential support that a prudent experimenter might wish to make. The theory will then be relatively useless, and it will certainly not suffice to achieve those purposes on which it was to concentrate. Yet this exclusive concentration on assessing the results of prudent experimenters was the only thing that entitled us to disregard the difficulties of constructing an adequately grounded measure for the relevance of a variable.

It follows that the set of relevant variables ought to have the same order in our support-theory as a prudent experimenter would assign to it, and obviously the safest way to ensure that this always happens is to require that the well-ordering relation be the same in both cases. But what well-ordering relation would a prudent experimenter adopt? Presumably he would seek to make his tests as revealing as possible, however small the number of variables involved. If conducting a test in regard to one variable alone he would select the variable that he believed most relevant, or at least as relevant as any of the others. If conducting a test in regard to just two variables, he would select the two that he believed more relevant than any of the others; and so on. So, ideally, the well-ordering relation for the set of relevant variables is that of greater relevance, so far as this can be judged. But where he believes two variables to be equally relevant or cannot judge which is more relevant, our experimenter will obviously have to fall back on one or more supplementary ordering relations, of which presumably the first to be invoked will be that of requiring less effort to manipulate.

For this purpose it does not matter if judgements of comparative relevance have a somewhat hazardous status. Admittedly the difficulty of devising a practicable and reliable way of measuring relevance prevents us, as was argued in §6, from assessing support in a way that takes degree of relevance into account. For the purpose of telling just how comparatively hazardous is a certain generalisation on the evidence of given test-results, we need a well-founded and reliable method of assessment. A hazardous method of assessing hazardousness is scarcely profitable. But we are not committed to such a hazardous method if we invoke judgements of comparative relevance as a basis merely for ordering the variables appropriate in a particular field. What is hazardous then is just the wise or prudent experimenter's judgement of how best to constitute his tests. Our method of assessing his test-results is not hazardous at all, because it does not need to take actual degree of relevance into account and is just geared to whatever ordering he himself adopts. Its correctness depends on correspondence with his own procedures, not on the truth of the tentative judgements of relevance that underlie these procedures.

It has still to be shown that all prudent experimenters in a particular field would normally agree in their tentative judgements of comparative relevance at any one time, and hence that the correct support-function for that field is uniquely determined. But this conclusion follows from plausible assumptions about the way in which tentative estimates of comparative relevance should be made. Since a variable v_i is in any case relevant to hypotheses of a particular type, if a variant of it V_i^1 has falsified one of these hypotheses in circumstances apparently similar to those in which another variant of it V_i^2 has failed to falsify the same hypothesis, it is plausible to suppose that variable v_i is to be tentatively judged more relevant than variable v_j if v_i has been observed to achieve such a feat in regard to more hypotheses than has v_j. This supposition will operate satisfactorily at any one time to determine the order of variables, and therewith the set of canonical tests and the correct support-function, for a particular field, so far as they can be known, provided certain further plausible assumptions can be made about prudent experimenters.

The first assumption is that they exchange information and always agree with one another about the falsificatory achievements of each variable. The second assumption is that they do not credit variable v_i with a higher falsificatory score than v_j just because v_i has been more often manipulated in tests than has v_j: the falsificatory score of a variable must be calculated as its ratio of falsifications to manipulations. The third assumption is that a variable's falsificatory score is not to be increased by relatively trivial means. In particular, if the record on which calculations are to be based includes the results of testing the hypothesis U_1, it must not also include the results of testing any other hypothesis U_2 that implies U_1, since falsification of U_1 implies falsification of U_2. The fourth assumption is that, in order to ensure some stability in experimental practice, there is some (expressly or tacitly) agreed margin by which v_j's score must come to exceed v_i's before the tentative order of priority between them is, tentatively, reversed. Finally, a fifth assumption is that if, despite the existence of such a margin, the order of variables is subject to somewhat frequent reversals, prudent experimenters will conjecture that materially dissimilar hypotheses are being treated as if they are materially similar. Their conjecture will be confirmed if division of their field into two or more distinct subject-matters succeeds in producing a corresponding number of comparatively stable hierarchies of tests.

Thus there is in principle no difficulty in programming a philosophically sophisticated support-function, of the kind discussed, for a particular field of experimental enquiry. In practice we may often know very little about what variables are relevant, or we may be very uncertain about the criteria of material similarity that will turn out to be fruitful. But in these circumstances, which occur especially in new fields, the philosophically sophisticated experimenter cannot help being just as doubtful as the philosophically unsophisticated one about how to compare E_1's support for U_1 with E_2's for U_2. It is a merit, rather than a defect, in a philosophical theory of inductive support that it should not ignore the way in which inductive assessment can be handicapped by such shortcomings in background information. So far as the programming of a support-function is to be subjected to empirical control, we are bound to be unable to design a well-substantiated programme if the requisite empirical data are still lacking. The best we can then do is to conjecture a programme and be prepared to revise our conjecture as further facts come to light.

But there are still a very large number of issues to be settled about such functions for ranking or grading inductive support—not least the problem of how to register unsuccessful test-results as well as successful ones.

§8. Some implications of support-grading

The proposed method of ranking or grading experimental support allows $s[U, E] > 0$ in some cases even where E contradicts U. A consequence principle now emerges for evidential propositions, and also a general conjunction principle. One possible objection to the latter is the problem created by evidence that appears to support two conflicting hypotheses. But there is a straightforward solution for this problem. The answer to another possible objection reveals that zero-grade support appears to spring indistinguishably both from unsuccessful test-results and from irrelevant or non-canonical evidence. But this paradox is avoided by noting the difference between unconditional dyadic assessments like '$s[H, E] = i/n$' and conditional monadic ones, like 'If E, then $s[H] = i/n$'. Further syntactic principles, for the logic of the monadic support-function, may now be developed.

If support is to be ranked or graded rather than measured, we are not faced with the problem of how to distribute quanta of significance between successful and unsuccessful trial-outcomes. We just have to think of U's support-grading, on the evidence given by E, in terms of how far U has remained unfalsified by the manipulation of cumulatively increasing combinations of relevant variables. We shall normally have $s[U, E] < i/n$ where E implies that U has failed test t_i or does not imply that U has passed any canonical test t_j where $j \geqslant i$; and we shall have $s[U, E] = 0/n$ where E implies that U has failed t_1, and also where E does not imply that U has passed any canonical test. It emerges quite straightforwardly that we can sometimes have $s[U, E] > 0/n$ even in cases where E contradicts U, which was a desideratum mentioned in §6 above. For example, if E implies that U has passed t_3 but failed t_4, we shall have $s[U, E] = 3/n$. E might do this in either of two ways. It might report a successful result under t_3, as described in §7, and an unsuccessful one under t_4. Or (and this possibility was not mentioned in §7) it might report a performance of t_i, where $i \geqslant 4$, such that U was falsified in all the trials in which one variant of v_3 was present (giving s[U, E] < 4/n) but positively instantiated in all the trials—the possible combinations of variants of the variables v_1 and v_2—in which some other variant of v_3 was present (which is tantamount to a successful result from t_3 and gives s[U, E] ⩾ 3/n). In this way, even when U is falsified by some variant of v_3—the variable manipulated in t_4 but not in t_3—we are still able to do justice to the extent to which U has, as it were, approximated the truth.

Indeed, one of the most striking ways in which support-functions differ from logical probabilities is that we do not always have s[U, E] = 0 where E is logically inconsistent with U, though in a theory of logical probabilities, we always have p[U, E] = 0 in such a case. Correspondingly, a theory of logical probabilities must give us p[U, E] = 1 where E logically implies U. But if E logically implies U we have no need to carry out an inductive assessment. The latter is only needed where deductive logic is of no avail. Hence the question how far E inductively supports U has normally to be determined in quite a different way from the question whether U is deducible from E. The scale of inductive support runs, not from logical contradiction to logical implication, but from no support by appropriate experimental tests to full support (see further §§19–22 below). An inductive support-function is not a measure of a natural scientist's success in behaving like a calculating-machine, as it were, but applies criteria derived from past experience to scoring his success at the prediction of future from present experience.

What are we to say, however, if E reports one occasion on which U has passed t_j and another occasion on which it has failed t_i where $j \geqslant i$ and $i < n$? It seems at first sight as though we are landed with the paradox of being committed to the truth of

(6) $s[U, E] \geqslant j/n \;\&\; s[U, E] < i/n$, where $j \geqslant i$,

even though E seems perfectly self-consistent. But in fact E would then describe a physical impossibility, if we accept the principle of 'same cause, same effect' and the consequential repeatability of test-results; and, if impossibilities are admitted as evidence, any hypothesis whatever may be supported.[1] Why is it impossible for such an E to be true, you ask? Well, if any combined manipulation of variables $v_1, v_2, \ldots, v_{i-1}$ sufficed to falsify U under t_i it would also have sufficed to falsify U under t_j, since t_j is a test in relation to all of the variables $v_1, v_2, \ldots, v_{i-1}, \ldots, v_{j-1}$. But if, as E reports, U was not falsified under t_j, it follows that $v_1, v_2, \ldots, v_{i-1}$ did not suffice to falsify U under t_i. The occasion on which U appeared to fail t_i must have been one where some variant or variants of another

[1] Cf. the formal metatheorem 472 in §22 below.

variable (or variables) occurred coextensively with the variant or variants in which U turned out to be false. That other variable (or variables) may be v_k, where $k > j$, or it may be a variable of hitherto unsuspected relevance, but it must bear at least part of the responsibility for U's reported failure on one occasion of testing under t_i, despite E's claim to the contrary—in the form of '$T_i x$'—when reporting the unsuccessful trial. In short, when E reports conflicting evidence of the type described, E cannot be true and we are not committed to (6) but only to $s[U, E] \geqslant j/n$.

The situation is not much different if we seem to have (6) where $i = j = n$. In such a case E reports conflicting results from two performances of the most thorough test t_n; and, though '$T_n x$' does not occur in E (cf. p. 56 above), these results are only intelligible on the assumption that some variable of hitherto undetected relevance is causing trouble. So, if the support-function itself is not to be rejected, with its reliance on the relevance of just $v_1, v_2, \ldots, v_{n-1}$, we must suppose E to state a physical or causal impossibility. The same cause should always produce the same effect.

Both types of case, it should now be noted, satisfy the syntactic principle

(7) For any E, F and H, if F is a consequence of E according to some non-contingent assumptions, such as laws of logic or mathematics, then $s[H, E] \geqslant s[H, F]$

since (7) holds where E and H are like E and U above and F is either that part of E which reports U's failure under test t_i or that part which reports its passing t_j. Moreover (7) holds for all other cases of H, E and F (where H is a universal proposition of the form 'Anything, if it is R, is S') according to the proposed method of ranking experimental support. It holds if E just implies two or more favourable test-results for H; it holds if E just implies two or more unfavourable test-results for H; it holds if E just implies H's success in one or more tests t_i and H's failure in one or more tests t_j, where $j > i$; and it holds if E also, or only, reports facts other than the results of canonical tests on H, since these can contribute nothing towards raising H's support above zero-level. So the principle holds too for substitution instances of first-order generalisations, in virtue of the uniformity principle—(23) of §4. (That principle is now sanctioned by the instantial comparability principle because our support-function is to have a finite, well-ordered set of values; cf. p. 33).

It is interesting to note also that (7), as a consequence principle for evidential propositions, produces the converse of the relation produced by the consequence principle for hypotheses that was discussed in §2.

Where F is a consequence of E we get $s[H, E] \geqslant s[H, F]$, but where I was a consequence of H we had $s[H, E] \leqslant s[I, E]$. So the equivalence principle, as given in §2, is implied by the conjunction of the two consequence principles.

Another syntactic principle that emerges from the proposed method of ranking support is a conjunction principle for first-order universal hypotheses, viz. (in the case of the dyadic support-functor)

(8) For any U_1, U_2 and E, if $s[U_1, E] \geqslant s[U_2, E]$, then $s[U_1 \& U_2, E] = s[U_2, E]$

where U_1 and U_2 are universal propositions. For, if E implies in relation to U_1 just that it is not adversely affected by cumulatively increasing combinations of relevant variables up to v_j, and in relation to U_2 just that it is not adversely affected up to v_i, where $j \geqslant i$, then clearly E implies in relation to the conjunction U_1 & U_2 just that it too is not adversely affected up to v_i; and, if E does not imply that U_2 passes some particular test t_k, it cannot be taken as implying that U_1 & U_2 passes that test. It follows from (8) that the conjunction of two universal hypotheses will always have the same level of support, on given evidence, as at least one of the two conjuncts. The conjunction will never have lower support than both conjuncts. This not only precludes support-functions for universal hypotheses from being logical probabilities (except in limiting cases), because of the multiplication-law for the probabilities of conjunctions. It also precludes them from being any functions of the logical probabilities concerned, because the argument developed in regard to the instantial conjunction principle in §3 applies at least as forcefully to the conjunction principle for universal hypotheses. The structure of the equation that obstructs mapping on to the probability calculus is the same in both cases, viz. $s[H \& I, E] = s[H, E]$. Indeed, the argument is perhaps even stronger in the case of universal hypotheses than in that of their substitution-instances. The argument is not conclusive where $p[I, H] = 1$, and so it is not conclusive for the substitution-instances of causal generalisations, as remarked in §3. But if H and I are universal propositions they will very often be logically and causally independent of one another. So there are no grounds for supposing $p[I, H] = 1$, and the anti-probabilist argument runs its course.

It is worth noticing also that with the aid of (8) we can establish, for normal E, a conjunction principle for any pair of singular statements, P_1 and P_2, that are substitution-instances of different first-order universal statements, U_1 and U_2, respectively, viz.:

(9) If $s[P_1, E] \geqslant s[P_2, E]$, then $s[P_1 \& P_2, E] = s[P_2, E]$

First, the instantial comparability condition—(3) of §2—gives us

(10) If $s[P_1, E] \geqslant s[P_2, E]$, then $s[U_1, E] \geqslant s[U_2, E]$.

Next, we use (8) and the equivalence principle—(1) of §2—to obtain

(11) If $s[U_1, E] \geqslant s[U_2, E]$, then $s[U_3, E] = s[U_2, E]$

where U_3 is a universal proposition that is equivalent to U_1 & U_2 in the way that e.g. '$(x)(y)((R_1x \rightarrow S_1x)$ & $(R_2y \rightarrow S_2y))$' is equivalent to '$(x)(R_1x \rightarrow S_1x)$ & $(y)(R_2y \rightarrow S_2y)$'. We then use the instantial comparability principle, as in deriving (5) of §2, to derive

(12) If $s[U_3, E] = s[U_2, E]$, then $s[P_1 \& P_2, E] = s[P_2, E]$,

and from (10), (11) and (12) we obtain (9). Moreover, since in virtue of the instantial conjunction principle—(4) of §2—(9) is also true for any P_1 and P_2 that are substitution-instances of the same universal proposition, we can now lay down, in the light of (8) and (9), and the uniformity principle (23) of §4, a quite general conjunction principle (in the case of the dyadic functor) for any H and I that are first-order universal propositions or substitution-instances of such, viz:

(13) If $s[H, E] \geqslant s[I, E]$, then $s[H \& I, E] = s[I, E]$

It may be objected to (8) that if two universal propositions U_1 and U_2 conflict with one another it is implausible that any proposition E should be said to support the conjunction U_1 and U_2 as highly as it supports one of the conjuncts. Obviously this objection deserves serious consideration. But presumably the conflict in question stems from its being either logically or physically impossible for the consequents of U_1 and U_2 to share a common instantiation in situations where their antecedents do share one, and in neither case can the objection be sustained.

In the first case (the logical conflict), it can be shown that on the proposed theory of support-ranking, if E gives more than zero-level support both to U_1 and to U_2, E's truth is impossible; and it should be no ground for worry, or even surprise, that impossible evidence supports conflicting hypotheses. The situation is similar to that discussed at the beginning of the present section, where E gave conflicting evidence in relation to a single hypothesis. For we are presuming here that U_1 has the form 'Anything, if it is R, is S_1' and U_2 has the form 'Anything, if it is R, is S_2' where 'x is S_1' is inconsistent, according to non-contingent assumptions, with 'x is S_2'. Now any report of U_1's passing a test t_i is necessarily also a report of a test-performance in which U_3 fails t_i, where U_3 has the form 'Anything if it is R is not S_1'. That is, if $s[U_1, E] = i/n$, where $i > 0$, then $s[U_3, E] < i/n$. But, since U_3 is a consequence, according to non-contingent assumptions, of U_2, the consequence principle for hypotheses (i.e. (2) of §2) entitles us to infer from this that if $s[U_1, E] = i/n$, where $i > 0$, then $s[U_2, E] < i/n$. Similarly, if $s[U_2, E] = j/n$, where $j > 0$, then $s[U_1, E] < j/n$. So if we assume that $s[U_1, E] = i/n$ and $s[U_2, E] = j/n$, where $i > 0 < j$, then we have to conclude both that $j < i$ and that $i < j$, which is impossible. Moreover, this contradiction springs directly out of an inconsistency in E, since (in the form of '$T_i x$') E reports the individuals tested in trials on U_1 to have been unaffected by other factors than the variants of variables $v_1, v_2, \ldots, v_{i-1}$, and (in the form of '$T_j x$') the individuals tested in trials on U_2 to have been unaffected by other factors than variants of $v_1, v_2, \ldots, v_{j-1}$, and yet in at least one of the two cases E also reports information inconsistent with this. In at least one of the two cases some other, hidden variable must have been at work, for the test-results to have occurred as reported. Otherwise, on the normal assumption that test-results are repeatable, the contradiction could not have arisen. Therefore E can only give more than zero-level support both to U_1 and U_2 when E's truth is impossible. It is quite possible for E to give positive support to U_1 and zero-level support to U_2, when U_1 and U_2 conflict. But then according to (8) E will give zero-level support also to the conjunction U_1 & U_2, which is unobjectionable.

In the second case we have to suppose that the joint operation of U_1 and U_2 is inhibited in a way that affects neither when operating alone. Perhaps, for example, swallowing a recommended dose of a barbiturate alleviates insomnia and swallowing a recommended dose of whisky alleviates depression, while swallowing both is fatal. In such a case, it is objected, the conjunction principle (8) cannot apply, because the conjunction of U_1 and U_2 must not be supposed to be as well supported as either conjunct alone. But in fact it would be a potentially dangerous error to suppose that rejection of (8) suffices to prevent trouble here. Even if you held only one of these two hypotheses to be well-supported, and believed accordingly just that, say, swallowing a barbiturate would alleviate insomnia, there would be nothing in your belief to exclude swilling the barbiturate down with whisky. The crux of the matter is that, if on good grounds you do believe it fatal to take whisky and barbiturates together, then what you must suppose to be well supported by the evidence is not just that if either of these is taken the corresponding malady is alleviated but that this happens if either is taken without the other. In other words a highly relevant variable for testing pharmacological hypotheses is constituted by what other drugs are taken at the same time. In order to escape falsification by trials that manipulate this variable and to be able to go on and achieve a high level of support, a hypothesis about the alleviative properties of barbiturates, or of whisky, must be suitably qualified in its formulation or suitably restricted in its domain of discourse. So rejection of (8) does not suffice to deal with the trouble at issue here.

Nor is it necessary to reject (8) for this purpose. Let us suppose that even in an unqualified or unrestricted form U_1 and U_2 still enjoy some support from available evidence. Then if we say that therefore the logically conjunctive proposition U_1 & U_2 also enjoys some support we are not saying anything more about simultaneous swallowings than was already implicit in the supposition that U_1 and U_2 both enjoy some support. Just as not every trial of U_1, or of U_2, tests what happens when other drugs are taken at the same time, so too not every trial of a universal hypothesis U_3, equivalent to U_1 & U_2, would test what happened when U_1 and U_2 were instantiated in the same individual and at the same time.

Another possible objection to (8) proceeds from the assumption

(14) For any E, H and I, if s[H, E] = s[I, E], then s[not-H, E] = s[not-I, E].

From (8) and (14), runs the objection, we may derive

(15) If $s[U_1, E] = s[U_2, E]$, then $s[\text{not-}U_1, E] = s[\text{not-}U_1 \ \& \ \text{not-}U_2, E]$.

Then by applying (14) to the consequent of (15) we derive

(16) If $s[U_1, E] = s[U_2, E]$, then $s[\text{not-not-}U_1, E] = s[\text{not-(not-}U_1 \ \& \ \text{not-}U_2), E]$

Finally, in virtue of standard logical equivalences, we derive from (16) and the equivalence principle, as given in §2,

(17) If $s[U_1, E] = s[U_2, E]$, then $s[U_1, E] = s[U_1\text{-or-}U_2, E]$.

But, the objector urges, it is easy to think of E, U_1 and U_2 for which (17) would be grossly paradoxical. For example, let U_1 be 'Any drug-addict is prone to mental instability', let U_2 be 'Any mentally unstable person is prone to drug addiction', and let E report evidence favourable to the existence of a correlation between drug-addiction and mental instability. Then presumably U_1-or-U_2 is better supported by E than is either U_1 or U_2 on its own, and this runs counter to what (17) implies.

But, though (17) is indeed an untenably paradoxical principle, the fault of generating it must be laid to the door of (14), not (8). If support-functions had precisely the same logical syntax as probability-functions, (14) would certainly be true, in virtue of the principle

(18) $p[\text{not-H}, E] = 1 - p[H, E]$.

But an objector to (8) is not entitled to assume here that support-functions are probability-functions, since this is part of what is expressly at issue. Nor does any analogue of (18) govern the proposed method of ranking experimental support. For, when E reports the results of non-canonical tests on H, or when E fails to describe the outcome of any appropriate trial of H at all—i.e. when E is quite immaterial to H—, we shall have both s[H, E] $= 0/n$ and s[not-H, E] $= 0/n$. Moreover, (14) is seen to be false when we consider any case in which E is quite irrelevant or immaterial to both H and not-H, but does give some support, and no opposition, to not-I (e.g. where I is 'Some R is not S', and so not-I is 'Anything, if it is R, is S'). In such a case we shall have s[H, E] = s[I, E] $= 0/n$, and also both s[not-H, E] $= 0/n$ and s[not-I, E] $> 0/n$, so that s[not-H, E] is not equal to s[not-I, E] as (14) requires it to be. It follows that the argument which generated the admittedly paradoxical principle (17) rests twice on the use of a false assumption (14), and the objection mooted is invalid.[1]

Nevertheless the need to answer this objection has brought to light a feature of the proposed support-ranking that may at first sight seem rather a flaw. We have s[U, E] $= 0/n$ not only when E reports that U has failed even the least thorough test, t_1, but also when E reports a non-canonical test-result or something quite immaterial to U. If it was a flaw in other theories to have no way of distinguishing adequately between the significance of successful trial-outcomes and the significance of unsuccessful ones, it would equally be a flaw in the present theory to have no way of distinguishing adequately between the zero support-value of unsuccessful test-results and the zero support-value of non-canonical tests or wholly immaterial evidence. However the latter distinction can in fact be drawn quite clearly, once an important ambiguity in every-day discourse about inductive support is fully recognised.

Up to now it has been assumed that sentences of the form 'E gives such-and-such support to H' may always be abbreviated with the aid of a dyadic support-functor. But to suppose that this is always so raises an awkward problem about the conditions under which we may infer to the

[1] It may be thought that (14) can be modified in a way that will sustain the objection, since the counter-argument considers only the case in which we have s[H, E] = s[I, E] = 0/n. Specifically, suppose we replace (14) by the principle

For all E, H and I, if s[H, E] = s[I, E] > 0,
then s[not–H, E] = s[not–I, E]

Now, this principle is normally true, since if E reports test-results supporting a universal proposition U it will normally not support not-U (cf. metatheorem 406 in §22 below). However, the objection under discussion needs to invoke such a principle twice—once to get (15) from (8) and once to get (16) from (15). In the first case it puts U_1 for H, and in the second not-U_1; and so if either s[U_1, E] = 0/n or s[not-U_1, E] = 0/n then on at least one of the two occasions the revised principle is not applicable.

conclusion that there does in fact exist *i*th grade support, according to the appropriate criteria, for H. It is certainly obvious that experimenters do sometimes feel themselves entitled to assess how much support there is for a certain hypothesis, and it is difficult to see what the point of testing hypotheses could be, if it were not to provide premisses from which, with the aid of appropriate additional premisses, such assessments could be inferred. But how can a proposition of the form '$s[H, E] = i/n$' constitute an additional premiss of this nature?

The solution commonly proffered for the problem is that the inference from E and $s[H, E] = i/n$ is legitimate if and only if E states all the available evidence. But this solution is doubly mistaken. In the first place it is quite inadequate as an account of reputable scientific practice. Experimenters are not thought rash just because they draw inferences from their own test-results without waiting to discover those of others. One performance of an appropriate test, carefully conducted, is often thought to provide sufficient evidence for assessing how much support exists for a particular hypothesis. Test-results are assumed to be repeatable, in normal circumstances, so no requirement of total evidence operates—no insistence on a record of all actual performances of the test. Secondly, the proffered solution elides the important fact that if a non-trivial conclusion of the form '$s[H] = i/n$' is required, and one premiss given is E, then it is logically necessary that the other premiss should be, or imply, 'If E is true, then $s[H] = i/n$'. So what have to be investigated, for the solution of the so-called 'detachment problem' here,[1] are the conditions under which such conditional monadic support-assessments are true and, in particular, the logical inter-relations between them and the corresponding unconditional dyadic assessments, like '$s[H, E] = i/n$'. It must be assumed that '$s[H, E] = i/n$' asserts the level of support that exists for H so far as we can tell just by looking at test-results reported in E, while 'If E is true, then $s[H] = i/n$' tells us what level of support exists for H if E is true. '$s[H, E] > 0/n$' asserts that there is some positive grade of support which E implies H to have, while 'If E is true, then $s[H] > 0/n$' asserts that E implies there is some positive grade of support

[1] It is worth while to distinguish between the detachment problems for dyadic and monadic inductive functions, respectively. The former concerns the logical conditions (stated formally in metatheorems 489 and 490 of §22 below) under which we can move from a dyadic assessment, of the form $s[H, E] \gtreqless i/n$, to a monadic one, of the form $s[H] \gtreqless i/n$. The latter concerns the logical conditions (stated formally in metatheorem 321 of §22) under which we can move from a monadic assessment, of the form $s[H] \geqslant i/n$, to a plain assertion of H. Both problems must also be carefully distinguished from the question of acceptability (cf. §1, p. 8 ff., and §10, p. 90 ff.). The old dispute about whether or not inductive support is 'relative to the evidence' was often cramped by the failure to draw these distinctions or even to distinguish clearly between a dyadic assessment of inductive support and a conditional monadic one.

for H: the scope of the existential quantifier 'there is some ...' is different in the two cases. But how do the several truth-criteria for conditional monadic support-assessments compare with those for unconditional dyadic ones?

If in relation to the variables $v_1, v_2, \ldots, v_{n-1}$ (where $v_1, v_2, \ldots, v_{n-1}$ is an ordered list of all the relevant variables in a particular field) E does not report a canonical test, we shall have $s[U, E] = 0/n$, since E tells us nothing significant about inductive support for U. But we need not then have, if E is true, $s[U] = 0/n$, since E, even if it contradicts or conflicts with U, need not exclude the existence (at some time and place in the universe) of some support for U or for a proposition implying U. To say that E does not support H is not at all the same as saying that E implies H to have no support. The former amounts to saying that there is no grade of positive support which, if E is true, H has: the latter amounts to saying that if E is true there is no grade of positive support which H has. The scope of the quantifier is again critical, and the sentence 'E gives 0th grade support to H' is unfortunately ambiguous between these two interpretations. Only if E implies consistently that U has failed test t_1, shall we have both $s[U, E] = 0/n$ and also, if E is true, $s[U] = 0/n$; and only if E implies consistently that U has failed t_i, shall we have both $s[U, E] < i/n$ and, if E is true, then $s[U] < i/n$.

Again, if E does not report that U, or some proposition implying U has passed any test t_j where $j \geqslant i$, then we shall have $s[U, E] < i/n$, but not necessarily, if E is true, $s[U] < i/n$, since the fact that E does not report such a test-result does not suffice to exclude it from occurring even if E is true. But if E reports that U or some proposition implying U has passed t_i, we shall have both $s[U, E] \geqslant i/n$ and also, if E is true, $s[U] \geqslant i/n$. Hence, if E reports consistently that U has passed t_i and failed t_{i+1}, we shall obviously have both $s[U, E] = i/n$ and also, if E is true, $s[U] = i/n$.

We can therefore lay down such syntactic principles as

$s[U, E] \geqslant i/n$ if and only if, if E is true, then $s[U] \geqslant i/n$

and

$s[U, E] \leqslant i/n$ if, if E is true, then $s[U] \leqslant i/n$.

Moreover, the difference between the zero support-value of E when E reports nothing but unsuccessful trial-outcomes, even for t_1, and the zero support-value of E when E reports non-canonical or immaterial evidence is now quite clear. In the former case we can assert both '$s[H, E] = 0/n$' and also 'If E is true, then $s[H] = 0/n$', while in the latter case we can assert both '$s[H, E] = 0/n$' and also 'E does not imply[1] that $s[H] = 0/n$'.

[1] On the nature of this (non-truth-functional) implication, cf. §20, p. 198 below.

The systematic deployment of the logical interconnections between monadic and dyadic support-functors is most conveniently left for formal treatment in §22 below (cf. metatheorems 482–499). But it is worth noting here that monadic support assessments conform to analogues of all the syntactic principles—except, of course, (7)—that have been stated for dyadic support-assessments in this and the preceding sections. These analogues are formed from the stated principles by deleting references to E or F, and changing expressions of the form '$s[H, E] \gtreqless s[I, E]$' to '$s[H] \gtreqless s[I]$'. It would be unnecessarily repetitious to construct an argument for each of these further principles. But two of them are worth stating explicitly here in order to establish informally a further principle from them that turns out to be of great fruitfulness when one comes to formalise the logical syntax of inductive support-functors.

The two to be stated here are a consequence principle for hypotheses (cf. (2) of §2), viz.

(19) For any H and I, if I is a consequence of H, according to some non-contingent assumptions, then $s[H] \leqslant s[I]$,

and a general conjunction principle (cf. (13) of the present section), viz.

(20) For any H and I, if $s[H] \geqslant s[I]$, then $s[H \,\&\, I] = s[I]$.

The principle to be established from them is

(21) For any H and I, if $s[H \rightarrow I] \geqslant i/n$,
then $s[H] \geqslant i/n$ only if $s[I] \geqslant i/n$

where '$H \rightarrow I$' designates indifferently any truth-functional conditional, equivalent to 'It is false that H is true and I is false'. For from (20) we obtain the closure of

(22) If $s[H \rightarrow I] \geqslant i/n$ and $s[H] \geqslant i/n$,
then $s[(H \rightarrow I) \,\&\, H] \geqslant i/n$,

and from (22) and (19) we obtain the closure of

(23) If $s[H \rightarrow I] \geqslant i/n$ and $s[H] \geqslant i/n$, then $s[I] \geqslant i/n$.

We can then use the formal-logical law of exportation to obtain (21) from the closure of (23).

Some corroboration for (23) may be achieved if we consider circumstances under which we might think the antecedent of (23) to be true. Suppose, for example, that we were dealing with the hypothesis

(24) $(x)((R_1 x \rightarrow R_2 x) \rightarrow (S_1 x \rightarrow S_2 x))$

and that we had at least ith grade support both for (24) and for

(25) $(x)(R_1 x \rightarrow R_2 x)$.

Then in accordance with the uniformity principle—(23) of §4—we shall have at least *i*th grade support both for any substitution-instance of (24) and also for any substitution-instance of (25). What (23) implies, in these circumstances, is that there will also be at least *i*th grade support for a substitution-instance of

(26) $(x)(S_1 x \rightarrow S_2 x)$

provided that all three substitution-instances are propositions about the same element. But this is exactly what we should expect to follow from the proposed method of grading inductive support in accordance with the results of canonical tests on universal hypotheses. Any report of a favourable outcome of a trial in test t_i on (24) would take the form of

(27) $(\exists x)(R_1 x \,\&\, R_2 x \,\&\, S_1 x \,\&\, S_2 x \,\&\, T_i\, x \,\&\, V_1^1 x \,\&\, V_2^1 x \,\&\, \ldots)$

and (27) would imply favourable outcomes for just the same trial on (25) and (26). So any typical conjunction of propositions like (27) that affords at least *i*th grade support for substitution-instances of (24) and (25) would afford the same for a substitution-instance of (26)—which corroborates (23). Indeed a test t_i on (24) may be viewed as a way of ascertaining whether, if there is *i*th grade support for any substitution-instance of (25), there is the same for any substitution-instance of (26). So a favourable test result tells us not only that if there is the one there is the other but also that there is *i*th grade support for this being the case. We can therefore assert not only (21) but also

For any H and I, if $s[H \rightarrow I] \geqslant i/n$,
then $s[s[H] \geqslant i/n$ only if $s[I] \geqslant i/n] \geqslant i/n$.

Beyond this point there is nothing to be gained by further informal statements of syntactic principles. A more than sufficient basis has already been stated for the systematic development of inductive syntax, and this development—with formal proofs—will be given in §22 below. But the philosophical defence of the principles already stated is far from complete. For first, we have so far considered universal hypotheses of only the very simplest form, as distinct from causal or quantitative ones or explanatory theories. Secondly, we have not examined whether our account of inductive support runs foul of any of the well known paradoxes, or riddles, of induction, in particular those due to the ingenuity of Hempel and of Goodman. Thirdly, we have yet to elucidate how the undeniably immense importance of probability-functions in experimental reasoning is nevertheless consistent with a radically non-probabilistic conception of inductive support. Fourthly, we have so far considered only the standard type of response to adverse experimental evidence, in which the hypothesis is conceived of as being refuted or undermined, rather than as being fully supported if appropriately reformulated. Fifthly, we have so far considered monadic and dyadic inductive functions only in relation to empirical propositions. In a somewhat wider perspective we may find that the syntactic principles already stated are more widely embedded in scientific and non-scientific discourse than has yet been shown, and that they have a correspondingly greater claim to the status of logical truth. Finally, we have argued much from the actual procedures of experimental science, without paying any heed to sceptical doubts about their rational justification. But these doubts will have to be dispelled or discounted before inductive syntax can confidently be reckoned a legitimate branch of formal logic.

III

Problems and Paradoxes about Experimental Support

§9. Support-grading for causal, correlational and theoretical hypotheses

The proposed method of support-ranking, with the conjunction principle that it entails, may be extended to cover causal hypotheses, by embodying trials on controls. It may also be extended to cover hypotheses stating one natural variable to be a certain function of another, and to deal adequately with the problems that arise through the existence of infinitely many different functions to fit a finite body of evidential data. It elucidates how these correlational generalisations can sometimes come to be unified within a single scientific theory whereby they come to corroborate one another. Indeed, ability to elucidate this phenomenon of consilience, as Whewell called it, must be taken as one of the criteria of adequacy for any philosophical elucidation of inductive support. But there is no instantial comparability principle that applies normally to correlational or theoretical generalisations.

We have been seeking a systematic method of ranking inductive support that can be taken to underlie familiar patterns of comparative assessment, where one hypothesis is said to better supported than another by the evidence of certain experimental tests or the same hypothesis is said to be better supported by one evidential report than by another. But we have so far considered the problem only in relation to universal hypotheses of a rather restricted type (and their consequences): viz. testable first-order descriptive generalisations of the form 'Anything, if it is R, is S' where 'R' and 'S' describe kinds, circumstances or characteristics of which the instances can, in principle, be observationally identified—with or without the aid of instruments. These were called 'empirical laws' by Mill,[1] and are the most elementary types of generalisation with which we are concerned. But they are elementary in virtue of their kind of meaning (only first-order, observational, etc.), not in virtue of their terms' being atomic or primitive, in any logically

[1] J. S. Mill, *System of Logic*, III, xvi. Mill also referred to some second-order generalisations as empirical laws.

important sense. Indeed, if the accepted criterion for the truth of 'x is R' or 'x is S' is the appearance of a certain reading on an instrument, then the acceptability of this criterion depends in turn—at least in part—on the strength of support that exists for the appropriate underlying theory or theories (e.g., in relation to metre-length steel measuring rods, for a theory about the rigidity, expansion, etc. of steel and for the geophysical theory underlying the metric scale).

The next step is to extend our account to causal hypotheses of the form

(1) Being R is the cause (or the sign) of any Q thing's becoming S.[1]

Such a hypothesis may be taken, for the purposes of inductive logic, to be equivalent to a conjunction of two elementary hypotheses, viz.

(2) Any thing-at-a-time, if it is Q and R, becomes S

and

(3) Any thing-at-a-time, if it is Q and not R, does not become S.

Perhaps, e.g., the hypothesis asserts that administration of such-or-such a drug is the only cure for a certain illness. Then all that is needed for test t_i on the hypothesis is a test t_i on each of the elementary hypotheses that compose the equivalent conjunction. In each combination of circumstances in which an experimenter attempts to falsify a hypothesis of the form of (2) he must also attempt to falsify a corresponding hypothesis about controls, that takes the form of (3). He must eliminate the possibility that, e.g., in clinical trials of a new drug, subjects' recoveries from their illnesses might be due to various natural causes or clinical circumstances rather than to their having been given the drug. So he must test for the existence of patients, or groups of patients, who are not given the drug but nevertheless recover to just the same extent. In other words, an experimenter has to use here something like what Mill called 'the method of difference' as well as what Mill called 'the method of agreement'. But on the present account (in contrast with Mill's) it is possible to rank the results of such tests.

[1] Normally, if we can discover an explanation of a thing's being both R and S that treats these as co-ordinate phenomena, then we say that R is the sign of S, whereas, if being S has to be explained by being R, and being R by something else, we say that R is the cause of S. But in discussing hypotheses like (1) here, I aim to show only that the method of assessing inductive support that was proposed in §§7–8 for the most elementary type of generalisation is in principle applicable to these hypotheses also. A detailed analysis of the concept or concepts of causation, or of explanation, is outside the scope of the present book.

The operation of causal generalisations is commonly assumed to be restricted to so-called 'normal circumstances'. This concept of 'normal circumstances', and some of the issues raised by it, will be discussed in §16 below.

Where U is the causal hypothesis (1) and there are $n-1$ relevant variables, we shall have $s[U,E] \geqslant i/n$ $(i > 0)$ if and only if E implies both (2) and (3) to have survived test t_i, and $s[U,E] \leqslant i/n$ $(i \geqslant 0)$ if and only if E does not imply both (2) and (3) to have survived t_{i+1}; and E will imply $s[U] \leqslant i/n$ $(i \geqslant 0)$ if and only if E implies that either (2) or (3) has failed t_{i+1}. So in effect the grades of support for (2) and (3) determine the grade of support for (1) in accordance with the conjunction principle for first-order universal hypotheses—(8) of §8. Moreover, that principle can now be seen to be applicable also to conjunctions of causal hypotheses like (1), for the same reasons as those given in §8 in relation to conjunctions of elementary hypotheses. (We have already seen, in §2, why the instantial comparability and instantial conjunction principles must apply to causal hypotheses.)

Not all causal hypotheses, however, assume that the phenomena with which they deal have a single cause. It may be hypothesised, for example, only that the administration of such-or-such a drug is at least one cure for a certain illness. Such a hypothesis has the form not of (1) but of

(4) Being R is a cause (or a sign) of a Q thing's becoming S.

Now (4) is equivalent, for the purposes of inductive logic, to a conjunction of (2) not with (3) but with

(5) There is some characteristic V, such that any thing-at-a-time, if it has V and is Q, is not R and does not become S.

For, if being R is only one of several characteristics that can cause a Q thing to become S, the merely negative characteristic of not being R does not suffice to prevent a Q from being S. What an experimenter needs here, for his controls, is some characteristic that is incompatible not only with being R but also with the presence of any other of the possible causes. But it cannot be assumed that he already knows all the other causes. So to test a hypothesis like (5) he needs to select as his control-type in turn one or more other variants $V_1, V_2, \ldots, V_m$ of the largest appropriate variable to which the characteristic of being R also belongs. He can then proceed as if testing the hypothesis

(6) Anything, if it has V_j and is Q, does not become S.

Where U is the causal hypothesis (4), we shall now have $s[U,E] \geqslant i/n$ $(i > 0)$ if and only if, for some such V_j, E implies both (2) and (6) to have survived t_i, and $s[U,E] \leqslant i/n$ $(i \geqslant 0)$ if and only if for any such V_j E does not imply both (2) and (6) to have survived t_{i+1}; and E will imply $s[U] \leqslant i/n$ $(i \geqslant 0)$ if and only if E implies that either (2) has failed t_{i+1} or, for any such V_j, (6) has failed t_{i+1}.

But how do we determine the largest appropriate variable from which to select variants as control-types for a particular causal hypothesis like (4)? The answer to this question is that experimenters must learn from experience here, just as in determining which variables are relevant to a certain class of materially similar propositions. Indeed we shall see shortly that there is a strong case for regarding this variable as being itself a kind of relevant variable.

A rather more involved problem presents itself if we try to extend our method of ranking experimental support to cover hypotheses about natural variables, such as the pressure of a gas or the velocity of a falling body. It is convenient again to consider only very simple hypotheses of the type in question, viz. hypotheses about the correlation between just two variables, whether quantitative or qualitative, which assume that the same individual can be ascribed variants of both variables at the same time. These hypotheses will be assumed to have the form

(7) For any characteristic R and any individual-at-a-time x, if x has R and R is a variant of variable v_{R}, then there is a characteristic S, such that x has S and S is a variant of variable v_{S} and $f(\mathrm{S}, \mathrm{R})$.

For example, v_{R} in (7) might be the time, measured in seconds, for which a body has been falling, v_{S} might be the velocity, measured in feet per second, at which a body moves, and f might be an appropriate dyadic second-order property (e.g., with '$f(\mathrm{S}, \mathrm{R})$' meaning that to have S is to be moving at a velocity, measured in feet per second, that is such-or-such a mathematical function of the period of fall, measured in seconds, to have been falling for which is to have R).

Clearly even the simplest correlational hypotheses, like (7), involve more than one order of logical quantification. They speak about more than just one domain—about variant characteristics as well as about individual objects. Correspondingly, tests on such hypotheses need to diversify their trials along more than one dimension. Test reports may still be written as conjunctions of existential propositions, each reporting the co-instantiation of the antecedent and consequent of (7) in some appropriate combination of variants of relevant variables (determined as before), with the appropriate safeguard '$\mathrm{T}_i x$' conjoined. But now two orders of existential quantification are required. Correspondingly, it will not only be necessary to manipulate one or more relevant variables that are variables not alluded to in the hypothesis in order to diversify the circumstances of the objects tested. It will also be necessary, in at least some tests, to manipulate the very variable, v_{R}, that is alluded to in the antecedent, or instead, where this v_{R} is a continuum, some finite set of non-overlapping intervals within v_{R}. For otherwise test-results might be due to special conditions prevailing at particular points on this parameter.[1] We have thus far to relax the initial prohibition (§5, p. 41) on there being any mention, in a hypothesis, of any variant of a relevant variable. I shall therefore speak of a variable as being inductively relevant to a testable hypothesis U if and only if either (i) it is inductively relevant to the particular set of materially similar propositions to which U belongs or (ii) it is an appropriately selected subset of a variable v_R such that either v_R or a member of this subset is mentioned in the antecedent of U. For example, it would not suffice to test a hypothesis about velocity of fall by manipulating solely the size or shape of the falling body, the medium in which it falls or its lateral proximity to large stationary objects. It would also be necessary to test such a hypothesis over a certain range of variation in the time for which the body has been falling.

But over what range? The trouble is, as has often been remarked, that however wide is the range of successful trial-outcomes in the available evidence for a particular quantitative correlational hypothesis, U, there often seems to be an infinity of other correlational hypotheses that also fit this evidence, but conflict both with U and with one another. To put the point in terms of (7): however many variants of v_R figure in the supporting evidence, each paired off with a variant of v_S, there often seems to be an infinity of different dyadic second-order properties, other than f, that map the second member of each pair on to the first. So how can any one of these rival hypotheses ever be better supported than any other? Moreover, the rival hypotheses thus conflict with one another in a modification of the sense that was discussed in the previous section (§8) in relation to hypotheses like 'Anything, if it is R, is S' and 'Anything, if it is R, is not S'. Such elementary hypotheses were said to conflict with one another if it was logically impossible for their consequents to share a common instantiation where their antecedents do share one. Rival correlational hypotheses, however, normally generate clashes of this nature at only some, not all, values of their antecedent variables. They can coincide over the values in some given array of evidence; and only

[1] It is assumed here, for clarity of exposition, that an experiment on a single object may determine the outcome of each trial. But this is by no means always the case in tests that involve the measurement of physical magnitudes: cf. §13 below, where the issue is discussed. It is also assumed here that simple correlational hypotheses can describe actually occurring phenomena. Perhaps, however, every correlational hypothesis that is well-supported involves some element of idealisation and may therefore be categorised as 'fact-correcting' rather than as 'fact-describing'. This concept of idealisation, and the close connections that exist between level of idealisation and level of inductive support, are discussed in §16 below: it is a serious mistake to suppose, as is sometimes supposed, that inductive logic is somehow refuted by the existence of 'fact-correcting' generalisations.

at some, not all, values is it then logically impossible for their consequents to share a common instantiation where their antecedents do share one. Correspondingly, in the type of conflict discussed previously it was not possible for the same evidential proposition, if self-consistent, to support even as many as two conflicting hypotheses, whereas in the present type of case it looks as though the same evidential proposition always supports an infinity of them. It thus seems rather more difficult to deny here that the proposed conjunction principle—(8) of §8—is able to produce the paradox of there being much positive support for the conjoint truth of conflicting generalisations.

The root of the problem lies in the interaction between the fact that a natural variable alluded to in a correlational hypothesis may have a continuum of values, or variants (as they are being called here), and the fact that no practicable test can embrace more than a finite number of trials. It looks as though those two facts force the conclusion that in such a case no practicable test could cover all the relevant varieties of characteristics an object might have, and hence that experimental support for correlational hypotheses cannot conform to the proposed syntax and semantics for inductive support-functions. But the two facts just mentioned do not really suffice to force such an unwelcome conclusion. This conclusion only follows if we assume that any mathematical function whatever may be invoked by f in (7), and therefore that in relation to any particular set of test-results we may have to deal with an infinity of materially similar hypotheses. Only then do the relevant varieties of characteristics that an object might have become infinite in number. But this assumption is quite unrealistic as an analysis of reputable scientific practice. Only certain, relatively simple functions are normally invoked. In effect, we can say, material similarity between hypotheses like (7) is to be defined not only in terms of the natural variables alluded to but also in terms of the types of connection asserted to hold between them—e.g. in terms of the types of mathematical functions invoked. Consequently, a finite and relatively small number of trials will normally serve to show which of any set of materially similar hypotheses is better supported even when all members of the set have some positive instantiations.

After all, in manipulating a relevant variable v so as to test whether a hypothesis of the most elementary kind is falsified by any of v's variants, we are in effect always determining which of certain rival hypotheses, such as 'Any R are S' and 'Any R are not S,' may be eliminated from serious consideration. Hence it is no special peculiarity of correlational questions in natural science, as distinct from more elementary ones, that the task of choosing between rival possible answers to them should be assisted by relatively restrictive criteria of material similarity. Correspondingly, as we relax these criteria and admit a greater variety of mathematical functions, say, in the formulation of hypotheses, so too we must widen the range of variation in their antecedent parameters over which hypotheses must be tested. In order to ensure the performance of trials at crucial values of those parameters there must be more variants in the relevant variable that is constituted by the set of intervals of v_R within which trials are to be performed.

This type of relevant variable thus differs from others in that alterations to the range of variants belonging to it can sometimes be justified on relatively *a priori* grounds. But it is for scientists themselves, not philosophers, to determine criteria for the admissibility of mathematical functions in particular fields. No doubt experimenters will in practice be influenced by such factors as computational facility and analogies with experience elsewhere. But we need not concern ourselves here with the precise form this influence should take, any more than with other practical issues about material similarity (cf. §5). In practice, no doubt, criteria for the admissibility of mathematical functions are determined implicitly, tacitly and loosely, rather than expressly and precisely. But we are entitled to assume that a prudent experimenter, who wishes to ensure comparability of test-results, will adopt sufficiently sharp criteria for this purpose and corresponding sets of intervals for the relevant variables involved.

Even on that assumption it might still seem possible for the conjunction principle to create trouble in certain circumstances. We can now guarantee that at least the most thorough test, t_n, will in principle include crucial trials for arbitrating between two hypotheses like (7) that allude to the same variables, v_R and v_S, but invoke different mathematical functions—giving v_S different values for certain crucial variants or values of v_R though not for all. But unless every canonical test so arbitrates, two such conflicting hypotheses could apparently both have positive support on the results of t_i, where $1 \leqslant i < n$; and then, according to the conjunction principle—(8) of §8—, there would apparently be some positive support for the conjoint truth of conflicting hypotheses. Nor can we rely on our assumptions of the form $T_i x$, as in (1), (2) and (3) of §7, to get us out of the difficulty—as they did in the case of conflicting elementary hypotheses. The trouble is that when we set out to justify such assumptions in testing correlational hypotheses, like (7) of the present section, we cannot hope to screen off any effect of a variant of v_R, since it is an effect of just this variable that our test is designed to reveal. It follows that every test of (7), from t_1 onwards, must include trials at critical values of v_R—i.e. at values that are crucial for arbitrating between hypotheses that conflict with (7) only in respect of the mathematical function they invoke. In other words, when we come to design the hierarchy of tests on a group of correlational hypotheses like (7), the relevant set of intervals in their antecedent variable, like v_R, must occupy first place in the well-ordering of relevant variables. Indeed it is scarcely implausible to suppose that a prudent experimenter would always regard this variable as the most relevant one for correlational hypotheses. He could hardly expect to make any progress at all towards learning whether the velocity of a falling body is such-or-such a function of the time it has been falling unless he deliberately set out to observe bodies that have been falling for appropriately differentiated periods of time.

There is in this respect a clear continuity of development as we proceed from the most elementary hypotheses, *via* causal hypotheses, to correlational ones. In each case the first test manipulates a variable to which attention is directed by the hypothesis' antecedent. In test t_1 on an elementary hypothesis only one variant of this variable is involved, viz. the variant specified in the antecedent itself. In test t_1 on a causal hypothesis at least two such variants are involved, viz. the variant specified in the antecedent and some other variant that acts as a control. And in test t_1 on a correlational hypothesis some larger number of variants of the antecedent variable is normally involved. Correspondingly we must now include this variable in the list of relevant variables and so speak of t_1 as manipulating a relevant variable, with the consequence that, when we come to take correlational generalisations into account, the number of different canonical tests and of distinct grades of support is to be reckoned equal to the number of relevant variables, rather than one greater than it (as was originally proposed, p. 53 f.). Indeed, from this point of view an elementary generalisation may be regarded as being equivalent to the limiting case of a correlational generalisation, where the variables correlated each have only one variant and accordingly the most relevant variable also has only one variant. Similarly a causal generalisation like (1) or (4) may be regarded as being equivalent to a (qualitative) correlational generalisation where the most relevant variable has two or more variants, viz. the putatively causal factor, described in the antecedent, and the appropriate types of control. So from this point of view, when Mill distinguished the method of difference, and the method of concomitant variations, from the method of agreement, he may be regarded as having been distinguishing the role of v_1 from the role of $v_2, \ldots v_n$, respectively, within any appropriately ordered set of variables $v_1, v_2, \ldots v_n$, that are the inductively relevant variables for some correlational generalisation.

It is evident that as we pass from elementary questions about variants to correlational questions about variables there is an important sense in which the domain of the enquiry tends to narrow. In a particular field of elementary enquiry there may be very many quite different questions for which answers can be proposed and tested, so that we can often be in a position to say that the answer to one question is better supported than the answer to another as well as that one answer to a particular question is better supported than another. There are very many species of birds, for example, and the morphology of each species is a possible topic of ornithological investigation. But when we pass to a field of correlational enquiry it is no longer the case that questions may be differentiated from one another by being concerned with different variants of the same variable, e.g. with different species of birds. Questions are now posed in terms of variables, not variants: e.g. how does the velocity of a falling body relate to the time for which it has been falling? Also, just as the questions themselves are to that extent more general and cover a wider range of phenomena, so too there are far fewer such questions to ask. Hence in regard to correlational hypotheses it is not implausible to suppose that a particular support-function may be concerned only with hypotheses about a particular natural variable and with the consequences of these hypotheses. We do not have to suppose that v_R, for example, is relevant to any other correlational hypotheses than those equivalent to propositions, like (7), that mention it in their antecedents. Even when our criteria of material similarity are thus relatively narrow we can still build up a list of the other relevant variables for hypotheses like (7) from experience with testing past answers to the relatively few questions now at issue; and there is still a place for comparative judgements between different answers to these questions or between the support given by different test-results to the same answers. But our criteria of material similarity *need not* be so narrow. It has therefore to be remembered that, while some $v_1, v_2, \ldots, v_n$ may be said to be inductively relevant to a particular correlational generalisation, only $v_2, v_3, \ldots, v_n$ in such a list may safely be said to be relevant to the particular set of materially similar hypotheses to which U belongs.

Moreover precisely the same support-function can apply not only to a group of materially similar correlational hypotheses, like (7), but also to every elementary generalisation that connects a variant of the natural variable with which they are primarily concerned to a variant of one of the other variables with which they deal—provided that we relax, as already suggested, the requirement (cf. §5, p. 41) that no variant of a relevant variable may occur in a hypothesis to be tested. For in testing such an elementary hypothesis, like

(8) Anything, if it has V_R, has V_S,

where V_R is a variant of v_R and V_S of v_S (e.g. 'Any body that has been falling for one second is moving at a speed of 16 feet per second'), we submit it to trial under every *possible* combination, with its antecedent, of variants of the appropriate relevant variables, and so no special difficulty arises about test t_1, as we now envisage it. In the case of an elementary generalisation, the antecedent itself describes a variant of the appropriate relevant variable v_1 (e.g. speed of fall), and is thus incompatible with descriptions of other variants of the same variable. Accordingly test t_1 is still constituted by just one trial.

Indeed we can see that the interconnections between support for elementary generalisations and support for correlational generalisations must be very close, once we notice that the arguments in favour of the consequence principles for hypotheses and evidential propositions (i.e. (2) of §2 and (7) of §8, respectively) are just as valid for correlational as for elementary generalisations. For, if an elementary generalisation H is such a consequence of a correlational generalisation U, then, according to the consequence principle for hypotheses, any evidence that supports U will give at least as much support to H, even if this evidence might otherwise seem quite immaterial to H. Conversely, according to the consequence principle for evidential propositions, the evidence for two or more elementary generalisations that are all special cases of U, as (8) is of (7), can give at least as much support to U as the corresponding report of a canonical test on U, since a merely first-order existential proposition

(9) $(\exists x)(V_R x \,\&\, V_S x)$

may imply the co-instantiation of the antecedent and consequent of (7).

We can thus take it as established that very much the same method of support-ranking applies to correlational generalisations as to elementary ones and that therefore the general conjunction principle (p. 64) with all its anti-probabilist implications, applies to the former as well as to the latter. But an instantial comparability principle does not normally apply to hypotheses, like (7), that generalise about kinds, characteristics or circumstances as well as about individuals. The basic reason why this is so is that, as we have seen, some part of the evidence may often fit both of two conflicting hypotheses of this type. I.e., a pair of such conflicting hypotheses may often have many pairs of substitution-instances that are mutually equivalent first-order generalisations. Now, consider two conflicting correlational generalisations U and U′ that are structured like (7) but invoke different mathematical functions in their consequents. We may suppose an evidential proposition E to report that U has survived test t_n, but that U′ has survived only one sub-set of trials out of all the trials which compose t_n, namely the trials under variant V_R. It follows that $s[U,E] = n/n$ and $s[U',E] < n/n$, and thus $s[U,E] > s[U',E]$. If so, then according to an instantial comparability principle we should have $s[P,E] > s[P',E]$ where P is any substitution-instance of U and P′ of U′. But U and U′ may have at least one pair of substitution-instances that are mutually equivalent first-order generalisations. If so, then according to the equivalence principle we may also have $s[P,E] = s[P',E]$. It is the possibility of generating such a contradiction, and a preference for retaining the equivalence principle at all levels, that prevents us from asserting an instantial comparability principle for higher-order universal generalisations, like (7), or a uniformity principle (analogous to (23) of §4) for them.

Nor can we assert an instantial conjunction principle for higher-order generalisations, since this would imply any two substitution-instances of the same higher-order generalisation to have the same grade of support on the same evidence; and a further antinomy might then be generated, as follows. Let H and H′ be two first-order generalisations about variants of the same pair of quantitative natural variables, such that H asserts anything that has V_R to have V_S and H′ asserts anything that has V_R' to have V_S'. Let E report successful test-results on H, and unsuccessful ones on H′. We should then have $s[H,E] > s[H',E]$. But there will always be some mathematical function enabling us to construct a higher-order correlational generalisation that has two substitution-instances which are equivalent to H and H′, respectively. So if substitution-instances of the same higher-order generalisation always have the same grade of support as one another on the same evidence we could also have $s[H,E] = s[H',E]$.[1]

The situation is interestingly similar when we come to consider the unification of different correlational hypotheses (appropriately modified perhaps) within relatively more comprehensive scientific theories. Here, too, it emerges, we may operate the same basic method of assessing inductive support in terms of test-results, and our assessments are therefore subject to the same conjunction principle for hypotheses. But here too no principle of instantial comparability, or instantial conjunction, applies.

A scientific theory is customarily regarded as a conjunction of several

[1] So far as causal hypotheses are concerned the situation differs in accordance with the form in which we consider them. The instantial comparability and conjunction principles apply to (1) because it is a first-order generalisation and has only singular propositions about individual things as its substitution-instances. But these principles do not apply to the second-order, correlational generalisation, equivalent to (1), that counts the first-order generalisations (2) and (3) among its substitution-instances, since (2) may well be better (or worse) supported than (3).

universal propositions rather than as a single generalisation like (7). These propositions constitute the postulates of a deductive system within which perhaps very many correlational hypotheses, and their substitution-instances, are derivable as theorems. The paradigm for such a theory, in modern science, was supplied by Newton's *Mathematical Principles of Natural Philosophy*, within which, for example, generalisations replacing Galileo's laws of falling bodies and Kepler's laws of planetary motion were derivable. But in order to see the essential similarity, as regards experimental tests and inductive support, between a scientific theory and a correlational hypothesis, we must conceive the former as a single generalisation, asserting that any correlation deducible from the postulates is correct. We must conceive a scientific theory not as a conjunction of postulates but as an equivalent generalised conditional of which the antecedent and consequent can be observably co-instantiated. Let $A_1, A_2, \ldots A_m$, be the postulates or axioms involved in such a theory. Then, at its simplest (and there will be an indefinitely large variety of more complex forms), the inductively testable equivalent will have the form of

(10) For any natural variables v_R and v_S, any dyadic second-order property f, any first-order characteristic R, and any individual element x, if x has R and R is a variant of v_R and the conjunction of $A_1, A_2, \ldots, A_m$ implies (logically or mathematically) that if x has R and R is a variant of v_R then there is an S such that x has S and S is a variant of v_S and f(S, R), then there is an S such that x has S and S is a variant of v_S and f(S, R).

For then, though (10) is a third-order generalisation, a merely first-order existential proposition, with the form of (9), may imply the co-instantiation of the antecedent and consequent of (10), as of (7), since V_R and V_S may be variants of variables that have a correlation deducible from the conjunction of $A_1, A_2, \ldots$ and A_m. The results of tests on elementary and correlational generalisations are thus available for the support of scientific theories. In practice a theory is tested by investigating such issues as whether in certain circumstances one dial always reads '25' (or within a specified interval of '25') when another reads '5'.

People sometimes speak as if correlational generalisations themselves (in addition to experimental data) can support a theory. Perhaps a stated value for the acceleration of falling bodies, for instance, may be said to support Newtonian mechanics. But in an analytical reconstruction of discourse about inductive support it is preferable to suppose that it is the evidence for a correlational generalisation which supports the theory, rather than that the generalisation itself supports the theory.

On the former supposition we can infer from the premiss that a report of such-and-such test-results is true, to the conclusion that there is such-and-such a grade of support for the theory. But if it is the generalisation itself which supports the theory we can only infer that conclusion from a premiss asserting the truth of the correlational generalisation, and so, since this truth is in turn a matter for inductive support rather than observational verification, we shall find ourselves entitled, by the experimental evidence, to infer only that there is support for there being support for the theory—which seems paradoxical. Accordingly, it will be assumed here not just that the results of experimental tests can sometimes support a scientific theory but also that, strictly speaking, they are the only type of support a theory can have.

Nevertheless, the construction of a theory must be capable of being understood as an attempt to unify and explain a certain set of correlational generalisations, and such an attempt has failed so far as the theory does not fit the working of one or more of these correlations over some range of values. Hence the set of correlations operating in a given field of enquiry determines, in effect, the set of inductively relevant variables for a theory in that field. More specifically, if U_1 is a correlational hypothesis about some natural variable in a certain field, and U_2 is such a hypothesis about a quite different variable in the field, and if T is a theory in that field, then one of the relevant variables for testing T will be the set of circumstances constituting the various trials of U_1 in the most thorough test of U_1 that exists, and another relevant variable will be the corresponding set of circumstances for U_2. It is no doubt largely because of this close connection between the structure of the various tests appropriate for T and the main different correlational generalisations which T should unify and explain, that people are often tempted to speak—loosely—as if these generalisations themselves, rather than their test-results, could support T.

Given that the operation of some natural variables is apparently more recalcitrant to theoretical explanation than that of others, we can assume that the relevant variables for T can, in principle, be serially ordered in terms of this apparent recalcitrance, just as the relevant variables for an elementary generalisation can be ordered in terms of the supposed relative frequencies with which their manipulation has falsified materially similar hypotheses (§7). On the basis of this, or any other, method of serial ordering a hierarchy of tests can be set up for T (and for any other theory about all the natural variables in the same field) wherein a cumulatively increasing number of relevant variables are manipulated, in just the same way as for elementary hypotheses, though it will often not be possible to combine a variant of one such variable with a variant of another. Consequently inductive support that is assessed in terms of the results of such tests will conform to whatever syntactic principles are determined by this method of assessment. In particular, for the reasons given already in §8, it will conform to the general conjunction principle and the consequence principle for evidential propositions.

But there is no instantial comparability, or uniformity, principle that applies generally to inductive support for theories like (10), because such a principle would produce an antinomy analogous to the one that has already been mentioned in regard to correlational generalisations. Consider two theories, T_1 and T_2, and a report of test-results, E, such that $s[T_1, E] > s[T_2, E]$. The trouble is that both T_1 and T_2 may have substitution-instances equivalent to the same correlational generalisation, U, i.e. the theories may offer alternative explanations of the same natural regularity. In that case a uniformity, or instantial comparability, principle would produce, with the help of the equivalence principle, the consequence that $s[T_1, E] = s[T_2, E]$, even though E in fact gave higher support to T_1 than to T_2. (Nor can an instantial conjunction principle apply to inductive assessments of theories, for reasons that are again quite analogous to those that apply in the case of correlational generalisations).

It is now possible to elucidate what Whewell called the 'consilience' of two correlational generalisations, U_1 and U_2, whereby one can come to corroborate the other when both are seen to fall under a single, unifying theory, though they both cover markedly different kinds of phenomena. Let us suppose that E_1 reports the successful result of the most thorough possible test on U_1, and E_2 on U_2. Let us suppose also that U_1 concerns phenomena that are thought to have been more recalcitrant to theoretical explanation than the phenomena with which U_2 is concerned. It will follow both that U_1 is open to corroboration by U_2, rather than U_2 by U_1, and also that the most relevant variable for a theory that unifies U_1 and U_2 is constituted by the set of circumstances of the trials reported in E_1. Hence we shall have $s[T, E_1] = 1/n$ and therefore, by the consequence principle for hypotheses—(2) of §2—, $s[U_1, E_1] \geqslant 1/n$. At the same time there is no reason to suppose, according to the support-function for hypotheses materially similar to T, that if E_1 is true there is higher than first-grade support for U_1. Hence we also have $s[U_1, E_1] \leqslant 1/n$. But $s[T, E_1 \& E_2] = 2/n$, and therefore, by the consequence principle again, $s[U_1, E_1 \& E_2] > s[U_1, E_1]$. The evidence for U_2 has thus come to support U_1 as well, and accordingly U_2 may be said to have corroborated U_1. And if we order the relevant variables for T differently, so that we have $s[U_2, E_2] = 1/n$ and $s[U_2, E_2 \& E_1] = 2/n$, then U_1 may be said to have corroborated U_2.

Of course, a merely trivial unification does not achieve any kind of corroboration, and it is clear, on the present analysis, why this is so. A trivial unification of U_1 and U_2 might be achieved by making T a theory of which U_1 and U_2 are themselves the postulates. But the terminology of such a theory would all belong to precisely the same category as that of U_1 and U_2, so there would be no ground for setting up a new support-function; and the general conjunction principle (p. 64) would ensure that T had just the same grade of support, according to the old support-function, as the less well-supported member of the pair, U_1 and U_2. To justify applying a new support-function the construction of T must introduce some radically new terminology, such that T is not just definitionally equivalent to a statable conjunction of correlational hypotheses formulated exclusively in the old terminology.[1] Otherwise there would be no new class of materially similar propositions. The mere conjunction, for example, of Kepler's laws of planetary motion with Galileo's laws about falling bodies does not constitute a case of consilience. What does constitute one is rather the construction of a theory of gravitational attraction within which both sets of laws are deducible along with indefinitely many other putative laws, such as, perhaps, those correlating the ebb and flow of the tides with the revolutions of the moon, and so on. To derive these other putative laws is to derive hitherto unknown ways of testing the theory, and is thus a way of filling out the list of its relevant variables.

Sometimes, in the very simplest type of consilience, just two or three natural variables may be seen to coincide with corresponding ranges of a single, more comprehensive variable. But even here the same requirement operates. Generalisations about the behaviour of ice, water and steam at different temperatures, for instance, can no doubt be unified

[1] I assume in such a case that the postulates of a theory generate not only theoretical propositions that are formulated in the new terminology but also bridging propositions that, when conjoined with these, enable us to derive propositions in the old (incorporated) terminology. But it is quite unnecessary, for the analysis of inductive syntax, to decide, for example, whether the bridging propositions are definitional or factual, i.e. whether the theoretical propositions should be interpreted as redescriptions of phenomena that were in fact described in the old terminology, or rather as descriptions of a reality that underlies such phenomena. There are, no doubt, many interesting and important issues that arise about the semantics of theoretical terms in natural science. But these issues are not germane to the problems of inductive logic. Nor do we need to suppose that the support-function appropriate to the old terminology alone has somehow been abolished or refuted by the construction of a new support-function that is appropriate both to the old and the new (theoretical) terminologies. It may still be useful to draw certain small-scale comparisons by reference to the old support-function, and no contradictions should result, in a genuine case of consilience. However, the treatment of consilience that I proposed in 'A Note on Consilience', *Brit. Jour. Phil. Sci.* xix (1968), p. 70 f. was mistaken, since, as shown in the present section, the instantial comparability principle does not normally apply to theories or to correlational generalisations.

into a theory about the behaviour of H_2O at different temperatures. But the latter is not then to be understood as a mere definitional abbreviation for the conjunction of the former: hypotheses about the chemical composition of a molecule of water are, in some contexts at least, open to test.

The substance of this point was grasped by both Bacon and Whewell. When we climb the ladder of axioms, as Bacon called them, the discovery of further axioms is vitally linked, Bacon thought, with the deduction of new particulars.[1] Similarly Whewell believed that, if consilience is to occur, it is necessary to achieve the 'superinduction' of some new conception over two widely differing sets of phenomena, and that within any theory that is to explain our existing knowledge some natural regularity must be deducible that has not been previously observed.[2] Where the present account seeks to add to the work of Bacon and Whewell is in its attempt to bring out the precise nature of the similarities and differences that exist between the structure of inductive assessment at the level of elementary hypotheses and its structure at higher levels.

No doubt there are very many features of scientific hypothesis-construction, and very many distinguishable types of scientific hypothesis, which have not been mentioned here. However our object, in the philosophical study of inductive support, is not to indulge in the natural history of scientific judgements, but to generalise about their logic on the basis of arguments that invoke appropriate criteria of adequacy. In particular we have just been arguing that the proposed method of assessing experimental support, and the syntactic principles that go with it—e.g. the conjunction principle for hypotheses—, can be adapted to fit certain types of causal, correlational and theoretical generalisations quite as well as they fit more elementary hypotheses. Indeed, the requirement that our syntax and semantics for support-functors should be adequate to the realities of experimental science commits us to showing that it can fit the more advanced areas of scientific generalisation as well as the more primitive ones. An explication of inductive support that cannot deal with paradigm hypotheses about causal connections or functional correlations, or that cannot elucidate consilience, is not to be taken seriously as a contribution to the philosophy of experimental science. But it would be tedious and unprofitable to discuss further types of scientific generalisation, and further adaptations of the method of assessing support, unless it can be shown that they present special problems for a theory of inductive syntax. A philosophical theory of inductive support has in any case to generalise over an unbounded domain, since it would be very rash for anyone to claim the ability to list

[1] F. Bacon, *Novum Organum*, I, ciii–cvi.

[2] W. Whewell, *The Philosophy of the Inductive Sciences* (1840), vol. II, p. 228 ff., esp. p. 242.

all future forms of scientific hypothesis. Correspondingly, in order to assess the merits of such a theory we must look to its success in dealing just with each major, structurally distinguishable, type of hypothesis for which some other theory of support, e.g. Carnap's, has notoriously been inadequate. In what follows, therefore, we shall not be primarily concerned to investigate yet further types of experimental hypothesis,[1] e.g. correlational generalisations that mention more than one variable in their antecedent, but to look instead at other dimensions of the inductive situation along which any general theory of inductive syntax has reason to fear difficulties.

§10. The comparability of support-gradings in different fields

It may be objected that belief in a plurality of support-functions is incompatible with the fact that we often think ourselves equally entitled to apply two widely different, and as yet ununified, scientific theories in the course of a single technological project. But this ignores the difference between acceptability and support. A dimension of universal assessment is provided by criteria for accepting a hypothesis at a particular time and for a particular purpose, even if the grades accorded by one support-function are incommensurable with those accorded by another. Further common forms of cross-field comparison are to be regarded not as comparisons of support under a single support-function but in various other ways.

Though support-functions in as yet unconnected fields of enquiry have a common logical syntax and a common method of assessing support by reference to a cumulative hierarchy of possible test-results, the actual test-hierarchies appropriate in each field are different because different lists of variables are relevant. Certainly, where there is already a unifying theory in a particular field, we can assume, on the present analysis, that an overarching support-function is available. But where there is no unifying theory a plurality of support-functions may be required. A critic may therefore object that certain important and immediate needs are not served, if the prospect of having a single, comprehensive support-function may be just as remote as that of having a single, unificatory theory for all hitherto explored fields of scientific enquiry. In particular, he may urge, we often need to apply generalisations from several quite different fields in a single technological project or even in the construction of a single piece of experimental apparatus. An internal combustion engine may rely just as much on the gravitational properties of the float

[1] However, §13 will discuss the structure of hypotheses involving estimates of magnitudes, and §16 will discuss the consequences of including an allusion to a relevant variable, or a variant of one, as a conjunctive component in the antecedent of a hypothesis to be tested.

in its carburettor as on the electrical properties of its sparking-plugs. A failure to unify the laws of gravity with those of electricity, or even to test them in the same way, did not deter the inventors of the internal combustion engine from judging that they knew enough about both sets of laws to be entitled to apply them with equal confidence in constructing a single, interconnected piece of machinery. Accordingly, the objection would run, a philosophical explication of inductive support that presents us with an as yet irreducible plurality of independent support-functions fails to supply certain criteria of scientific judgement that in practice we all assume ourselves to have already.

The premiss of this objection is quite correct. The proposed explication of inductive support does not permit serious comparisons, at all points of the scale, between the grade of support available for U_1 and that available for U_2, if U_1 and U_2 are materially dissimilar. No appropriate test on U_2 would be canonical for U_1's support-function. Only if we could measure relevance would the situation be seriously different. Each test-result could then be weighted for the relevance of the variables manipulated in it. A single, comprehensive support-function would suffice, and all test-results could be compared with one another in respect of the quantity of support they afforded for the hypotheses tested. But since no adequate measure of relevance seems possible (§6) we are left with a plurality of independent support-functions.

Nevertheless, though the objector's premiss is quite correct, his conclusion does not follow. The conclusion is invalid because it wrongly identifies judgements of acceptability (i.e. of acceptance-worthiness) with judgements of inductive support. For a particular technological project, whether it be building a motor-car engine or curing a disease, what we require to determine is whether the scientific generalisations we need to apply—the ones that constitute answers to the various questions involved—are sufficiently acceptable, at that time, for the purpose involved. Mere judgements of inductive support are not enough, for at least three reasons.

First, even if the inductive support for a certain hypothesis is assessed on the basis of all evidence already obtained, the loss to be suffered from relying on it for a particular purpose if it turns out false may be so great that more thorough tests need to be carried out before any decision can be made to accept the hypothesis for that purpose. Before being launched on the open market a new medical drug has often to be submitted to clinical trials on selected human patients even if it has already shown itself satisfactory in preclinical trials on experimental animals. In such a case it is as if each relevant variable has to be systematically manipulated before a hypothesis can be considered sufficiently well supported to deserve acceptance. Indeed there are circumstances when even a fully supported hypothesis—and one that constitutes a direct answer to the question asked—may not be acceptable as a solution to a particular problem. Admittedly, to say that a hypothesis is fully supported is to imply that it is true (though we may well be wrong to believe that we know all the relevant variables for testing it, and so we may be wrong to believe that it is fully supported). But sometimes—cf. §16 below—a hypothesis achieves full support (or is thought to achieve it) only at the cost of being very complex and riddled with qualifications that guard against unsuccessful trial-outcomes. Where this kind of complexity is excessive there is an intellectual loss that makes even a fully supported hypothesis unacceptable. In other cases, however, and in particular where there is great gain to be achieved if the hypothesis turns out correct and little or no loss if it turns out incorrect, a hypothesis may deserve acceptance on rather less evidence than that which only the most thorough test appropriate to it could afford.

Secondly, a hypothesis that deserves accepting for a particular purpose at one time may come not to do so, not only through changes in the gain-or-loss situation or in the amount of evidence obtained, but also through the discovery that the application of some other hypothesis may serve the same purpose more cheaply, say, or on the basis of better supporting evidence. So in deciding to accept one hypothesis for a particular technological purpose one is also, at least implicitly, rejecting the use of any other and perhaps also rejecting the worthwhileness of investigating whether any other such is to be discovered.

Thirdly, we have to bear in mind that judgements of inductive support on given evidence, whether conditional and monadic in form or unconditional and dyadic, are themselves empirically refutable, and some of them may be better or worse supported than others. If a hypothesis that is taken to be fully supported turns out to be false, or two evidential reports seem to justify conflicting support-assessments, then either some of the experimental test-results are being wrongly interpreted or there is something wrong with the support-function for hypotheses of that kind. The thalidomide tragedy showed that even test-procedures which are highly regarded at one time may later be thought insufficiently thorough. There is always the risk of a hidden variable—a variable of as yet unknown relevance—like the factor of pregnancy or non-pregnancy. There is also the risk of a superfluous variable—a variable which we manipulate in our tests in the belief that it is relevant, which would add corresponding support for a hypothesis in accordance with the appropriate support-function, but which in fact is not relevant at all because no variant of it is ever really responsible on any occasion for falsifying any materially similar hypothesis. For either reason it may be false to claim that s[U, E] $= i/n$ or that if E is true s[U] $= i/n$, even though the claim accords with hitherto accepted criteria for grading hypotheses materially similar to U. So all the above considerations, about balancing the gain-or-loss situation against the grade of evidential support at present obtained, apply also, at a higher level, to the grade of support for the inductive assessments on which we might think to base our acceptances or rejections. In a relatively new field of enquiry we may be much more reluctant than in an older one to accept an assessment of a hypothesis that is based on test-results under all variables which are believed relevant.[1] But a

[1] The generalisation that any variable is relevant to a particular set of materially similar propositions if it belongs to a certain set of variables is regarded here as being fully supported if the manipulation of each such variable has falsified some of these propositions, in the way specified in §5, and an ordering of this set is regarded as being acceptable if it is that which accords with the beliefs of a competent experimenter about order of relevance. A critic might therefore object that this cannot be treated as a case of inductive support, in the normal sense, since nothing apparently needs to be done to test whether each variable is equally relevant for each of any distinguishable sub-species among these materially similar propositions. But the situation here is similar to one of the type discussed in §16 below, where a single trial outcome is shown to be capable of giving full inductive support to a sufficiently qualified version of a hypothesis. In the present case such a qualification is to be taken as implicit in the notion of material similarity. We are concerned with relevance for a particular set of propositions only so far as the members of this set are sufficiently similar to one another for relevance to them to be unaffected by whatever differences also exist between them. But experimenters may, of course (cf. §5), come to have reasons for changing their minds about which hypotheses are materially similar to which. It might also be objected that in practice we rarely try to establish systematically, by testing hypotheses in a given field under the manipulation of widely different types of variable, that there is support for the proposition that *only* such-and-such variables are relevant in that field. Nevertheless we have to distinguish between testing a hypothesis successfully in relation to every variable that is at present believed relevant, and testing it successfully in relation to what are at present believed to be all the relevant variables; and the latter belief is just as much capable of enjoying inductive support as the belief, say, that only such-and-such factors cause plant diseases.

On the regress involved here to higher and higher orders of assessment, cf. §20, p. 202 f. below. Someone who feels strongly that ultimate certainty is never justifiable in science may object on those grounds to a concept of full inductive support, or inductive establishment, which implies that if a proposition is fully supported it is true. But, while sympathetic to the premise of this objection, I would deny that it constitutes any ground for objecting to the present analysis of inductive support. I construe the premise as claiming that, for any scientific hypothesis H, either no detectably true proposition (i.e. no proposition of which the truth may be determined by a finite set of observations) fully supports H, or no such proposition fully supports any proposition that a particular detectably true proposition fully supports H, or no such proposition fully supports any proposition that a particular detectably true proposition fully supports a proposition that a particular detectably true proposition fully supports H, and so on *ad inf.*: somewhere along this infinite regress a hypothesis has to be accepted on less than full support if any hypothesis is accepted at all. In other words, for any scientific hypothesis H, either we can never know that H is true, or we can know that H is true but can never know that we know, or we can know that we know but can never know that we know that we know, and so on. No doubt this thesis is correct, if, as I have argued, assertions of inductive support in experimental science are themselves empirical hypotheses. But whether or not it is

counterweight to this reluctance may be supplied by the prospect of the benefits that would be derived from accepting the assessment if it were true.

It is natural to compare this situation with what Carnap called 'the rule of maximising the estimated utility'.[1] It looks as if U's degree of acceptability in a given situation is mainly a function of how well U is inductively supported by the experimental evidence, how directly and simply U answers the questions at issue, and how useful U's truth would be. But we cannot in fact construct a parameter of acceptability here,[2] because grade of inductive support is not merely not a probability or function of probabilities—it is not even a measurable quantity of any other kind (cf. §6 above).

This is a far less neat and exact situation than we might prefer to have, and could indeed have if inductive support were measurable. Nevertheless it suffices to answer the objection that if there is a plurality of support-functions we have no dimension of assessment within which generalisations from quite different fields may be judged equally worth applying in a single technological project. Patently the overall, common standards that we invoke when deciding which hypotheses to rely on, in building a complicated piece of machinery, are universal standards of acceptability, not separate hierarchies of tests. Acceptability, indeed, is just as much in point, for this kind of purpose, if we want to rely on the evidential value of several different types of instrument-reading in tests on a scientific hypothesis, as it is if we want to rely on an aircraft's carrying human passengers safely. The acceptability of a hypothesis may be judged in relation to intellectual projects as well as practical ones. In particular, a judgement of comparative acceptability is all that is rationally possible when, in times of scientific revolution, there is a conflict not only between rival scientific theories, but also between rival criteria for the assessment of a theory's evidential support.

But, a critic may rejoin, there is more support for Boyle's law than for the proposition that penicillin cures every normal case of septicaemia, or more support for the proposition that penicillin cures every normal case of septicaemia than for the proposition that worms can acquire the memory of other worms by eating them, or more support for the proposition that the sun rises every morning in the east than for the proposition that albino dwarfs are unhappy. Yet these examples are all from widely different fields of enquiry, and have nothing to do with acceptability.

correct the thesis itself employs, or presupposes, a concept of full support. Hence the thesis affords no conceivable ground for objecting to the view that some such concept has a place in any analytical reconstruction of inductive reasoning.

[1] *Logical Foundations of Probability*, p. 264 ff.

[2] For the situation in regard to statistical probabilities, cf. §§12–13 below.

So how can such comparisons be explicated unless under the aegis of a single, comprehensive support-function?

But what this rejoinder points to is rather the need to make yet further distinctions in the ways we commonly talk about support. For in addition to the simple assertion, implicitly under the aegis of the same support-function, that one hypothesis has more support than another, we can also assert that the support enjoyed by one hypothesis, which is well supported under its own support-function, is of a higher type than that enjoyed by another hypothesis, which is well supported under a different support-function. The one hypothesis, perhaps, is the consequence of a well-supported scientific theory, as Boyle's law follows from the kinetic theory of gases, while the other hypothesis is not as yet incorporated into any scientific theory. Since any generalisation achieves a new source of support when it is so incorporated (§9), a distinction between (at least) higher and lower types of support here is quite legitimate. So, though Boyle's law cannot strictly be said to have *more* support than the proposition about penicillin, according to our proposed explication of inductive support, it certainly has support of a higher type.

Again, though intermediate grades of support under different support-functions are incommensurable with one another (for the reasons discussed in §§5–6), there is clearly a sense in which any hypothesis that has zero-grade support according to its own appropriate support-function—whether through actual test-failure or through the non-existence of any canonical test-results—has the same standing as any other, and a standing inferior to any hypothesis that has some positive grade of support according to its own support-function. So though people may sometimes say that the above proposition about penicillin is better supported than the one about worms, what they could just as well say is that the former has some properly inductive support by appropriate standards while the latter has none. Similarly any hypothesis that has full, i.e. top-grade, support according to some appropriate support-function has the same standing as any other and a standing superior to any hypothesis that has less than full support according to any support-function. For if a hypothesis really has top-grade support according to some appropriate support-function it cannot be more strongly established. In other words, if there are any laws of physics, say, or of ornithology, they are all laws of nature.

It is tempting, too, to suppose that, if the available evidence refutes neither of two generalisations but includes a very much larger number of instances in the case of one hypothesis—like the proposition about the sun—than in the case of another, then the former is better supported than the latter, however wide apart are the two fields of enquiry. But this 'enumerative induction', as it is often called, is not to be confused with inductive support in the sense discussed here. As will be argued in §12 below, it is rather to be regarded as an estimate of a statistical probability, and to confuse it with inductive support leads to paradoxes of the kind discussed in the next section, §11.

If, finally, there are yet other cross-field forms of comparison in common use, one can only suppose that they are based on untestable assumptions about equalities or inequalities of relevance and therefore do not merit elucidation in any experimentally oriented philosophy of inductive support. Of course, it would be possible to shun monadic and dyadic functors of the kind already described, in favour of a new kind of dyadic or triadic functor, respectively, where the additional argument-place would be filled by a reference to an ordered set of variables. Inductive assessments would take the form $s[H,\{v_1, v_2, \ldots v_n\}] = i$, or $s[H,E,\{v_1, v_n, \ldots v_n\}] = i$; and a typical assessment would state that (on the evidence of E) H has ith grade support on the assumption that v_1, $v_2, \ldots$ and v_n are all and only the inductively relevant variables. A certain type of commensurability would then emerge. But such assessments would be *a priori* (or analytic) in every case. They would fail to represent the empirical corrigibility of beliefs about hypotheses' grade of inductive support. Correspondingly, the commensurability that they permitted would be relatively superficial. From an *a priori* truth of the form

$$s[H, E, \{v_1, v_2, \ldots v_n\} > s[H', E, \{v'_1, v'_2, \ldots v'_m\}]$$

we should certainly not be entitled to infer the judgement that E in fact gives more inductive support to H than to H′, since this judgement is, normally, an empirically corrigible one.

§11. The elimination of Hempel's and Goodman's paradoxes

Not all the generalisations in a particular set of materially similar propositions are to be regarded as testable hypotheses. In particular, a testable hypothesis must be seriously liable to falsification by the causal operation of at least one relevant variable. E.g., 'Anything, if it is a hare, is grey' is testable, but not 'Anything, if it is not grey, is not a hare'. Thus Hempel's paradox is dissolved by insisting that equivalents are inter-substitutable in statements about inductive support, but not in statements about favourable test-results. Nicod's criterion, with its ambiguous use of the verb 'to confirm', obscures the point. Similarly, Goodman's 'grue' paradox disappears as soon as we replace Nicod's criterion by the thesis that inductive support accrues to a hypothesis H *via* favourable test-results on H itself or on hypotheses bearing certain logical relations to H.

It is evident that experimental tests of the kind described cannot be applied to every member of a particular set of materially similar propositions. They cannot, for example, be applied to propositions that are not conditional in form. Indeed we specified initially (§5) that only testable universal propositions, in a sense of 'testable' still to be defined, could be treated in this way. But how is that sense to be defined, and how does this definition affect the nature of inductive tests? Three conditions seem necessary.

Clearly, first, we require experimentally testable propositions to be generalised conditionals since we have to investigate whether they are falsified by any one of such-and-such combinations of circumstances, and we cannot report on this, for a universal proposition U, unless we first identify an object as being one of those about which U asserts something, e.g. unless we first identify an object as being R when U asserts that anything which is R is also S. It may seem that U could have the form $(x)(Sx)$, i.e. 'Anything is S'. But that is only because some assumption about the domain of discourse is here implicitly supplying an antecedent for Sx.

Secondly, we require testable propositions to be those for which both favourable and unfavourable test-results are genuinely and non-trivially possible. Neither they, nor the antecedents or consequents of their substitution-instances, may be propositions that are logically true or logically false, or propositions that are logical consequences or contradictions of accepted mathematical postulates[1] or of other non-contingent assumptions. To observe the satisfaction of the hypothesis' antecedent, and the satisfaction or non-satisfaction of its consequent, must require serious experimental effort. Nor may a testable proposition be one of a pair of mutually equivalent propositions U_1 and U_2 such that any reports of favourable test-results for U_1 imply the falsity of U_2. No genuine test is possible in such cases, because we cannot treat favourable test-results as being results that eliminate rival hypotheses: what is in fact eliminated there is not a rival, but an equivalent. Moreover, just as logically true propositions are regarded as not testable because they are trivially satisfied by anything whatever, so too some kinds of evidence must be regarded as instantiating certain kinds of generalised conditional

[1] I assume that some mathematical postulates, e.g. those for elementary arithmetic, are accepted *a priori* in any experimental investigation, whatever attitude is adopted towards other systems, such as Euclidean geometry. Of course, we all have to live with the fact that, as Church and Turing have shown, there can be no comprehensive decision procedure for the consistency of a proposition, either in logic or in mathematics. It follows that we may conceivably be unable in practice to discover whether a certain proposition is testable or not. But it is most unlikely that such a proposition would ever be of the slightest interest to any experimenter.

inadequately, because the antecedent or consequent of the conditional is satisfied trivially. Specifically, a conditional proposition is not to be regarded for these purposes as positively instantiated when its antecedent is not satisfied; and so too neither the antecedent nor the consequent, if itself a conditional or the conjunction of several conditionals, is to be regarded as satisfied on such trivial grounds, but only when the antecedent and consequent of each subordinate conditional are both satisfied. Similarly, if the antecedent of the main conditional is a disjunction, so that we have a generalisation of the form

Anything, if it is either R_1 or R_2, is S,

then twice the normal number of trials will be needed. If such a generalisation is to be adequately instantiated for the purposes of an inductive test, we shall need, in each of the combinations of circumstances, V, that are varied in the test, not only something that is R_1 and S and has V, but also something that is R_2 and S and has V; and a corresponding requirement will apply where the antecedent of a subordinate conditional is disjunctive in form. All trivial forms of positive instantiation are inadequate as experimental evidence.

Thirdly, just as the observation of supporting evidence must not be made trivially easy because of the logical structure of the hypothesis tested, so too it must not be made too easy in relation to the relevant variables; and this restriction has at least two consequences.

If the support-function were one appropriate to dealing with a scientific theory, it might be altogether too easy for a single elementary hypothesis in the same field to pass all the appropriate tests, because many of the trial circumstances would be impossible to combine with the satisfaction of the hypothesis' antecedent circumstances. So where theories are grouped together with correlational generalisations and elementary propositions, as a particular set of materially similar propositions, it is the theories, not the other propositions, that are testable.

On the other hand, even if the support-function is appropriate just to elementary propositions, there is the trouble that any universally quantified conditional has infinitely many equivalents and positive instantiations may be very much easier to find for one equivalent than for another, because their occurrence may be very much less inhibited by the manipulation of relevant variables. Correspondingly, falsifications may be more difficult to find if we operate with relevant variables on the antecedent of one of these equivalents, than if we operate on another. For example, variation of circumstance from summer to winter may prevent occurrence of positive instantiations for

(11) Anything, if it is a hare, is grey

but not for its contrapositive equivalent

(12) Anything, if it is not grey, is not a hare.

There is no shortage, whether in summer or in winter, of brown dogs, black cats or white mice, though grey hares are scarce or non-existent in a northern winter. But the point is not that very many more things are not grey or not hares than are grey or hares, respectively. This may well be so. But, even if it were not, the proper hypothesis to test would still be the one about hares rather than the one about not-grey things. Nor should it be thought that the absence of negatives in (11) and their presence in (12) are what makes the difference. Sometimes the logically more complex equivalent, containing more negatives, is the testable proposition. For example, it is much easier, under normal manipulation of relevant variables, to find positive instantiations for

(13) Anything, if it is a baby and grows healthy bones, is fed with milk

than to find them for the equivalent

(14) Anything, if it is a baby and is not fed with milk, does not grow healthy bones;

and correspondingly it is (14), not (13), that we should regard as testable. What makes the difference is not the numbers of individuals that satisfy the antecedent or consequent, nor yet the numbers of negatives present in them, but rather the precise effect, or direction of operation, of the relevant variables. Alteration of season makes some animals change colour: it does not make some coloured things change the animal species to which they belong. Addition of calcium to a baby's milk-free diet prevents rickets, but giving calcium to healthy babies does not prevent them from drinking milk.

So, as another condition for a member of a particular set of materially similar elementary propositions to be testable, we must require something like the following. Let the proposition in question be 'Anything, if it is R, is S'. Then 'R' and 'S' must describe variants of some variables v_R and v_S, respectively, such that 'V_R' and 'V_S, describe other variants of v_R and v_S, respectively, and manipulation of at least one relevant variable sometimes prevents things that are V_R from being V_S. A prudent experimenter, who wishes to test his hypotheses as severely as possible, will always prefer to relate his experiments to propositions which satisfy this condition. His hypotheses must be seriously liable to falsification by the relevant variables.

But it is important to note here that the domain of inductive support is very much wider than that of experimental tests. Very many non-testable propositions can have positive support. If there are favourable test-results for U, for example, then on that evidence any proposition H that is a logical or mathematical equivalent of U, has positive support. Also, on that evidence, if U is a first-order generalisation, then any substitution-instance of U itself, or of any of U's equivalents, has the same grade of support as U—by (23) of §4. Also, if U implies or is implied by H we can determine lower or upper limits, respectively, on H's grade of support, in accordance with the consequence principle for hypotheses. Moreover, for simplicity of systematisation, we shall find it convenient to suppose that all logically true propositions have maximum grade support. Indeed, one reason why the logical syntax of inductive functors is so important is that in these ways and in others it enables us to derive grades of support for non-testable propositions.

The failure to distinguish between positive support, on the one hand, and favourable test-results, on the other, is one source of the puzzlement about Hempel's paradoxes. Essentially what Hempel pointed out[1] is that, if anything confirming a hypothesis H is also supposed to confirm all H's equivalents, then any universally quantified conditional has some equivalents to which it is paradoxical to apply a certain commonly accepted criterion of confirmation. According to this criterion, which will here be referred to as Nicod's criterion,[2] an object, or ordered couple of objects, etc. confirms a universal conditional hypothesis if it satisfies both the antecedent and the consequent of the hypothesis. So that, for example, a white mouse confirms, paradoxically, the hypothesis (11), since a white mouse satisfies both the antecedent and the consequent of the equivalent hypothesis. (12).

Moreover, contrapositives like (12) are not the only equivalents that give trouble, if Nicod's criterion of confirmation is accepted. As Hempel pointed out, any hypothesis of the form

(15) $(x)(\mathrm{R}x \rightarrow \mathrm{S}x)$

has an equivalent of the form

(16) $(x)((\mathrm{R}x \,\&\, -\mathrm{S}x) \rightarrow (\mathrm{R}x \,\&\, -\mathrm{R}x))$.

Yet no object whatever can satisfy both the antecedent and the consequent of (16), since the consequent is self-contradictory. So it looks as though, according to Nicod's criterion, no object whatever can confirm

[1] C. G. Hempel, 'Studies in the Logic of Confirmation', *Mind* liv (1945), p. 1 ff. and p. 97 ff., reprinted in C. G. Hempel, *Aspects of Scientific Explanation* (1965), p. 3 ff.

[2] Actually it is a generalisation of Nicod's criterion, cf. Hempel, *Aspects of Scientific Explanation*, p. 11.

(16) even though very many may confirm its equivalent (15). Or again, the hypothesis

(17) $(x)(y)(-(\mathrm{R}xy \,\&\, \mathrm{R}yx) \rightarrow (\mathrm{R}xy \,\&\, -\mathrm{R}yx))$

is confirmed, according to Nicod's criterion, by the couple a,b, if we suppose that Rab and $-$Rba. But (17) can be shown to be equivalent to

(18) $(x)(y)(\mathrm{R}xy)$

which is flatly falsified by the very same couple a,b, since we have supposed that $-$Rba.

Nicod's criterion of confirmation is by no means the same as that proposed above (§7) for what is to count as a favourable result of a canonical test. Specifically, Nicod's criterion does not state that other circumstances—different combinations of variants of relevant variables—may also have to be present. Yet some such circumstances have to be present in every trial outcome that is to be deemed favourable, other than in test t_1, when an object satisfies both the antecedent and the consequent of the generalisation under test. Nor does Nicod's criterion mention insulation from the effects of variants of certain variables. Yet some such insulation has to be present throughout every canonical test-result other than a result of test t_n. Nevertheless, without the testability restriction—i.e. if any universal hypothesis whatever were regarded as testable—just the same troubles would arise for our method of grading inductive support as Hempel showed to arise for Nicod's criterion of confirmation. First, the finding of white mice in winter as well as in summer would seem to give support to the hypothesis that anything, if it is a hare, is grey. Secondly, such a hypothesis would also have an equivalent, like (16), that seemed quite incapable of support from the results of experimental tests. Thirdly, precisely the same test-results would seem to be favourable for some hypotheses, like (17), despite being unfavourable for their equivalents, like (18).

But all these paradoxes are averted by the restriction on testability. First, for the reasons already explained, (11) gets no support from the finding of white mice under any manipulation of the normally relevant variables. Secondly, (16) is not testable, because of its self-contradictory consequent, but it nevertheless has the same grade of support, on any evidence, as (15). Thirdly, neither (17) nor (18) are testable, just because test-results of the kind suggested can favour one while being unfavourable to the other.[1]

[1] It may be noted that no paradox arises for universal conditionals like $(x)(y)(\mathrm{S}xy \rightarrow \mathrm{R}xy)$, as distinct from universal categoricals like (18). If we suppose the truth of Sab & Rab & $-$Rba, the couple a,b instantiates positively both $(x)(y)(\mathrm{S}xy \rightarrow \mathrm{R}xy)$ and also its equivalent $(x)(y)(\mathrm{S}xy \rightarrow (-(\mathrm{R}xy \,\&\, \mathrm{R}yx) \rightarrow (\mathrm{R}xy \,\&\, -\mathrm{R}yx)))$.

In effect what we are doing here is to insist on a difference in the level of referential opacity, or non-extensionality, between the scope of the term 'support' and the scope of the term 'test'. Within neither scope can we substitute truth-functionally equivalent propositions—propositions that share the same truth-value—for one another without sometimes affecting the truth-value of the statement being made. But in assessments of support, whether monadic or dyadic, we can substitute any logically or mathematically (or otherwise non-contingently) equivalent propositions for one another without ever affecting that truth-value, whereas we cannot do even this within the scope of the term 'test'. That is because, when we think of evidential propositions as reporting the satisfaction of the antecedent, and satisfaction or non-satisfaction of the consequent, of tested hypotheses, any constraints we place on the concept of evidence tend to generate corresponding constraints on which hypotheses are to be tested. It would be different if, *per impossibile*, a proposition asserting the conjoint satisfaction of the antecedent and consequent of U_1 were always equivalent to a proposition asserting this of U_2, whenever U_1 and U_2 were mutually equivalent. For then the level of non-extensionality generated by 'test' would have to be the same as that generated by 'support'. But since this equivalence does not always hold we must, as it were, be more careful about what hypotheses we state to have been tested successfully than about what hypotheses we state to have been positively supported.

Thus the equivalence principle applies only to statements about support and not to statements about test-results. If U_1 and U_2 are equivalent, then support for U_1 is always also support for U_2, but a test on U_1 is not always also a test on U_2. One source of the puzzlement about Hempel's paradoxes is the obfuscation of this important fact by the way in which use of the verb 'to confirm' blurs the distinction between positive support and favourable test-results. But another, equally important, source is the conception of favourable evidence for a universal conditional as being constituted just by joint satisfaction of its antecedent and consequent. For it is only when one accepts that such a satisfaction, if it is to count at all, must have been achieved under some concurrent manipulation of relevant variables, that one can see how the causal direction in which these variables operate often makes a test on U_1 quite valueless, whatever its results, when a test on its equivalent U_2 may be well worth while. Of course, there is a notion of so-called 'enumerative induction', according to which a bare joint satisfaction of antecedent and consequent, without any variation of relevant circumstances, gives some support, however small, to a conditional generalisation. But we shall see in the next section, §12, that when this so-called enumerative induction is properly understood, as an estimate of a probability, Hempel's paradoxes disappear there for other reasons. The paradoxes are only puzzling until the vague and ambiguous term 'confirmation', has been analysed out into its various possible meanings of 'support', 'favourable test-results', and 'sample-based estimation', and the different implications of these meanings have been articulated.

Much the same is true also of another group of paradoxes, which Goodman has formulated in terms of the same verb 'to confirm'.[1] Suppose that the predicate 'grue' applies to all things examined before a certain moment *m* if and only if they are green and to other things if and only if they are blue. What Goodman pointed out is that, if all emeralds examined before *m* are green, then at *m* the evidence seems to confirm equally both 'All emeralds are green' and 'All emeralds are grue'.[2] Consequently, incompatible predictions about emeralds after *m* seem to be equally well confirmed by the same evidence. Moreover, it is particularly paradoxical if the evidence that all emeralds examined before *m* are green confirms a prediction that emeralds after *m* are blue just as much as it confirms the prediction that all emeralds after *m* are green. For in this way, it seems, any evidence can come to confirm any prediction whatever, and, if so, evidential confirmation can have no value at all as a guide to beliefs about the future. For example, suppose 'grue' applies to all things examined before a certain moment *m* if and only if they are green and to other things if and only if they exist in a world in which pigs have wings. Then, if emeralds examined before *m* are green, we can predict that after *m* pigs will have wings. Nor is the type of paradox confined to temporal extrapolations from observable evidence. If 'grue' is redefined so as to apply to all things examined in the experimenter's own laboratory if and only if they have one characteristic, and to other things if and only if they have a different one, then an analogous argument seems to lead to the absurd conclusion that no experimenter is ever entitled to draw any conclusions about the world outside his laboratory from what goes on inside it.

The first point to notice here is that the paradox hinges essentially on the existence of such a curious predicate as 'grue'. For, if we spell out the generalisation 'All emeralds are grue' in terms of the definition for 'grue',

[1] N. Goodman, *Fact, Fiction and Forecast* (1954), p. 74 ff.

[2] Goodman's paradox is made even more mystifying if a token-reflexive expression like 'now' is put in place of '*m*' in the definiens of 'grue'. For then the sentence 'All emeralds are grue' expresses a different proposition each time it is uttered, and evidence that confirms one such proposition, according to Nicod's criterion, may nevertheless disconfirm another. However, Goodman's inductive paradox about Nicod's criterion may conveniently be kept apart from the rather more general problem of token-reflexive expressions and the so-called 'fugitive' truth-values they seem to generate.

the paradox dissolves very quickly. We then have a generalisation of the form

(19) $(x)(Rx \rightarrow ((S_1x \leftrightarrow Mx) \& (S_2x \leftrightarrow -Mx)))$

where 'R' stands for 'is an emerald', 'S_1' for 'is green', 'M' for 'is examined before moment *m*' and 'S_2' for 'is blue'. Now (19) is equivalent to the conjunction

(20) $(x)((Rx \& S_1x) \rightarrow Mx) \& (x)((Rx \& Mx) \rightarrow S_1x) \&$
$(x)((Rx \& S_2x) \rightarrow -Mx) \& (x)((Rx \& -Mx) \rightarrow S_2x).$

So, by the equivalence principle, (19) must have the same grade of support on given evidence as (20) and therefore, by the conjunction principle (i.e. (8) of §8) the same grade of support on given evidence as the least well supported conjunct in (20). But, if the evidence consists only of emeralds that are green before time *m*, then, even if it had been gained under appropriate manipulation of relevant variables, the evidence could not give any support at all to the third and fourth conjunct in (20) though it might support the first and second quite well. It could not constitute test-results for any hypotheses that implied the third and fourth conjuncts, though it might do so for the first and second. Hence, on the evidence in question, (20) has no support at all.

If, however, the consequent of (19) is replaced by the single expression 'S_3x' standing for 'is grue', so as to make the hypothesis in question not (19) but just

(21) $(x)(Rx \rightarrow S_3x)$,

the argument via (20) and the conjunction principle might well fail to get off the ground. This would certainly happen if the category of non-logical terms to which 'grue' belongs failed to include terms like 'is before moment *m*'. For then no conjunctive equivalent of (21) would be available for appraisal under the same support-function.

Goodman's own solution of the problem thus presented is to propose criteria by which, in effect, neither (21) nor any equivalent of (21) would be a testable hypothesis. For him, no hypothesis is projectible, i.e. confirmable, if it conflicts with a hypothesis which projects a much better entrenched predicate, i.e. a predicate that has been projected much more often. So (21) is not projectible because it conflicts with 'All emeralds are green' and 'green' is much better entrenched than 'grue'. But this solution of the paradox is quite unacceptable. It would block the path of enquiry, if ever a weakly entrenched predicate like 'grue' were really needed as a rival to some strongly entrenched one like 'green'. Perhaps there are (as yet unknown) discontinuities, turning-points, or crucial dividing-lines, in some fields of enquiry, so that things of a certain kind have the queer characteristic of exhibiting one feature on one side of the line, and another on the other. A philosopher of science is not entitled to solve his logical problems by proposing criteria of confirmability that would, in effect, anticipate the future of science. Indeed we already have many highly reputable scientific generalisations that explicitly refer to particular regions of space or time. Kepler's laws of planetary motion are one example, and palaeontological generalisations about, say, the mesozoic era are another.

But, once it is accepted that no evidence is capable of giving positive inductive support to a hypothesis unless it is constituted by favourable test-results under manipulation of at least the most relevant variable, even the 'grue' formulation, as in (21), ceases to generate any paradox. Every canonical test on a testable equivalent of (21) must either include at least one trial before *m* and at least one trial after *m* or be insulated from the effect of this variable. For presumably no variable can be more relevant to a hypothesis about a discontinuity or dividing-line than the variable that includes both sides of the line. So if there were any clear and consistent evidence at all in support of 'All emeralds are grue', it would be evidence against 'All emeralds are green'. Also, it would be evidence of which a part would have been obtained after moment *m*. Any evidence that has all been obtained before moment *m* cannot give even 1st grade support to 'All emeralds are grue' because it cannot have been obtained with concurrent manipulation of the most relevant variable or under insulation from its effects.

Another way to arrive at this conclusion is to notice that 'All emeralds are grue' is what I have earlier (in §9) called a (qualitative) correlational generalisation, though it tends to masquerade as an elementary generalisation like 'All emeralds are green'. For in fact it correlates the emerald-examined-before-*m*/emerald-examined-after-*m* variable with the green/blue variable. Like any other hypothesised correlation, therefore, this one too must always be tested over suitable variations in its antecedent variable.[1]

In sum, Goodman's paradox arises only through the mistaken acceptance of Nicod's criterion as a sufficient condition of inductive support.

[1] This point does not depend at all on regarding 'green' as being somehow a more primitive term than 'grue'. The situation is just the same if 'green' and 'blue' are regarded as non-primitive terms. 'Green' might then be defined as 'grue if and only if examined before *m* and bleen if and only if examined after *m*' (where 'bleen' applies to emeralds examined before *m* just in case they are blue and to other emeralds just in case they are green), and 'blue' would be defined correspondingly. But the correlational generalisation 'All emeralds are grue' will still have no support in fact; while, if 'All emeralds are green' is regarded as a generalisation that correlates the emerald-examined-before-*m*/emerald-examined-after-*m* variable with the grue/bleen variable, it will be well supported.

That criterion almost fits the least thorough test, t_1, for elementary hypotheses (cf. §7, p. 54), and so there is a certain superficial plausibility in applying the criterion to 'All emeralds are grue'. But the criterion does not fit any test for correlational hypotheses, and 'All emeralds are grue' is in fact a correlational hypothesis.

No doubt Nicod's criterion may instead be treated as a touchstone for so-called enumerative induction. But, when the latter notion is adequately elucidated, the paradox will turn out to disappear in that case for other reasons (cf. the next section, §12). The paradox can only remain puzzling if the difference between this notion, on the one side, and inductive support, on the other, is obscured by discussion of the topic in terms of Nicod's criterion and the verb 'to confirm'.

IV

Induction and Probability

§12. Induction by simple enumeration

What Bacon and Mill called 'induction by simple enumeration' is more properly to be regarded as a special case of the estimate of a certain kind of magnitude, viz. the estimate that a particular probability is 1 or very nearly 1. The measure of such an estimate on the basis of given evidence is always the (higher-order) probability that a sample of the size in evidence is one in which the magnitude in question matches the actual magnitude for the statistical population as a whole, within a stated degree of approximation. But in a statistical procedure for accepting or rejecting hypotheses this measure is normally employed in the context of some relatively *a priori* decision: to operate with such-and-such confidence intervals, say, or with such-and-such significance levels. If enumerative induction is regarded thus, as the estimate of a probability, neither Hempel's nor Goodman's paradoxes can arise. Also, it becomes clear that so-called enumerative induction cannot establish conclusions from which an important type of counterfactual conditional is deducible. Such conclusions must enjoy support from the results of experimental tests in relation to relevant variables.

As a criterion for justifying acceptance of a generalisation people sometimes invoke the sheer number of its positive instantiations, in the absence of any falsification. The circumstances of these instances do not need to have been selected or contrived on the grounds of their inductive relevance. But the more numerous the instances are, the better is supposed to be the evidential justification for accepting the hypothesis. Bacon and Mill called this mode of argument 'induction by simple enumeration', and viewed it with some reserve.[1] Enumerative induction

[1] E.g. F. Bacon, *Novum Organum*, I, cv; and J. S. Mill, *System of Logic*, III, xxi, 2–3. It may be argued that if a generalisation has been positively instantiated in a large number of cases, and negatively instantiated in none, this makes it likely that the generalisation has been positively instantiated in a wide variety of relevant circumstances; and thus it may be objected that support by enumerative induction is just a cruder form of the inductive support already considered in §§7–11. But it is always rash to infer relevant variety from mere multiplicity, however great the multiplicity, unless there are adequate additional grounds for the inference; and if there are adequate additional grounds, we have a case to which we can, in principle, apply the consequence principle for evidential propositions—(7) of §8—not a case of enumerative induction.

was for them very much a second-best in ordinary science, when compared with induction by variation of circumstance; and while they had much of value to say about the latter they contributed little to the study of the former. Indeed the very title they gave to enumerative induction tended to obscure its structure. For by calling it a method of induction they suggested that it should be classed along with the establishment of universal hypotheses by experimental variation of circumstance; and they thus dissociated it from arguments to conclusions that could be represented only by probabilities of less than 1.

This point is worth emphasising a little. Induction by experimental variation of circumstance is for at least two reasons primarily focused on the establishment of support for fully universal hypotheses. First, experimental tests on a hypothesis are always assumed to be, in principle, repeatable. You ought to be able to get my results in your laboratory, and I to get yours in mine. If a certain combination of experimental circumstances once causes something, it always does—cf. §8. So the logical form of an experimentally testable hypothesis must not impose any restriction on test-repeatability, even if the terms of the hypothesis (extinct animals, rare diseases) sometimes do. The hypothesis must be fully universal in form. Secondly, as we have also seen (§9), experimental variation of relevant circumstances may be regarded as a method of eliminating rival hypotheses; and, since each successful trial-outcome eliminates at least one such rival, the hypotheses in rivalry with one another must be universal. They must be conceived to state that anything, if it is R, is S, not just that most things, or a certain percentage, have this characteristic.

On the other hand, arguments from sheer numbers are not intrinsically tied to the variation of relevant circumstances and thus to the establishment of support for fully universal hypotheses. One sample of a certain size is no better and no worse than another, as a basis for argument, so far as numbers alone are to count. Admittedly argument from sample size is often combined in practice with argument from the results of relevant variations in the method of selecting samples (cf. §13 below). But if argument from sample-size is considered abstractly, and on its own, it makes no causal or quasi-causal claims about the falsificatory efficacy or inefficacy of different factors and so is not committed to some form of repeatability, as are reports of test-results. Nor does this method of argument seek, implicitly or explicitly, to eliminate rival hypotheses, as does the manipulation of relevant variables. Correspondingly arguments from samples need not be primarily directed towards conclusions that are fully universal in character. Such arguments can be just as well concerned to argue from the fact that 80% of observed R's have been S's to the conclusion that the probability of an R's being an S is not much different from 4/5, as to argue from the fact that all observed R's have been S's to the conclusion that R's are fairly certain to be S's. In other words they characteristically aim to describe a population collectively, not distributively.[1] What counts for merit here is the size of an evidential sample, not the thoroughness of an experimental test, and this type of merit can be attained irrespective of whether the argument is about a uniformity or about a ratio.

If, therefore, we regard so-called enumerative induction as being concerned to establish probabilities of approximately 1, we can conceive it as just a special case of a wider form of argument, viz. the estimation of probabilities from samples. Moreover, samples can be used as a basis for estimates of other magnitudes too, besides probabilities. We can also use a sample to estimate the mean value, for example, of a certain variable in a statistical population, or the median, or the variance, or the standard deviation. So what the older philosophers misleadingly classified as the inferior of two inductive methods can now be viewed in a quite different perspective. It is just a special case of one kind of magnitude-estimation. The ugly duckling of the older philosophers has grown up, as it were, into one of the many swans in the aviary of modern statistics.

The structure of magnitude-estimation has been extensively explored by mathematicians over the past forty years; and, though violent

[1] An attempt is sometimes made—e.g. by C. S. Peirce, *Collected Papers*, ed. C. Hartshorne and P. Weiss (1932), 2.664, and K. R. Popper, 'The Propensity Interpretation of the Calculus of Probability and the Quantum Theory', in *Observation and Interpretation*, ed. S. Körner (1957), p. 65 ff.—to predicate probabilities distributively, rather than collectively, by trying to conceive them as a kind of propensity, or dispositional property, of individuals. But is the 50% probability of falling heads, say, supposed to belong to this coin-toss or to this coin? If the former, the predication is genuinely distributive but it cannot be justified by analogy with dispositional predication. In order to be able to ascribe a disposition, like fragility, we normally need to know something about the conditions that bring about its display, whereas we commonly resort to statements of probabilities just because our causal knowledge is (at least as yet) deficient. Even if an adequate explication of modern particle physics did require us to contrive a method of interpreting the probability calculus that assigns probabilities to individual events, it would be no help to conceive such a probability as a disposition or propensity. On the other hand, if the 50% probability of falling heads is supposed to belong to this coin, rather than to this coin-toss, the predication is collective rather than distributive because something is supposed assertable about the population of the coin's tosses. Nothing is gained by speaking of this probability as a propensity, since the coin's propensity is just to fall heads, or to fall tails, *simpliciter*. The probability measures that propensity and is not to be identified with it, any more than the length of a needle is identical with the needle. Popper has now made it clear, in 'Quantum Mechanics without "The Observer"', *Quantum Theory and Reality*, ed. M. Bunge (1967), p. 33 ff., that he wishes to regard a quantum-theoretical probability as a property of a type of experiment, not of an individual experiment. But he does not make it clear what he now supposes to be gained by regarding it as a propensity of the experiment-type, rather than as a frequency in the long-run population of experiments of that type.

controversy has raged over some issues, an underlying core of common practice has established itself, in science, commerce, industry and administration, in respect of the issues with which we are at present concerned. The validity of this practice is rooted in a mathematical fact of which the importance was apparently first discerned by Jacques Bernoulli. In a normal, finite population, for any given size of sample, a bigger proportion of such samples is, within any given interval of approximation, representative of the actual magnitude in the population than is representative of any other magnitude. As the size of sample increases, so does, for any given interval of approximation, the proportion of representative samples to non-representative ones, and, for any designated proportion of representative samples, a reduction occurs in the interval within which each representative sample can approximate the actual magnitude. Also, by speaking of probabilities rather than proportions we can generalise this structure of mathematical relationships so as to cover infinite populations. In other words, if, for instance, the probability of an R's being an S is $p \pm \epsilon$, where ϵ is some relatively small interval, then as the size of sample increases, so does the probability that within a sample of R's the relative frequency of S's will be $p \pm \epsilon$.

Where the magnitude to be estimated is itself a probability, we are in effect concerned here with two probabilities, the probability to be estimated, on the one hand, and, on the other, the probability that the relative frequency in a sample of a certain size is within a certain interval of this magnitude. The latter probability may also be regarded as the probability that an estimate in accordance with a sample of such-or-such a size is within a certain interval of success. But whereas the probability to be estimated is expressed by a probability-function that takes sets as its arguments, the estimate itself is assessed by a probability-function that takes propositions of certain types as arguments. For, if the probability to be estimated is the probability that an R is an S, it may be written as p[S, R]. But the probability of such an estimate's being successful, within an interval of approximation ϵ, if the estimate is based on a relative frequency r_n in an n-membered sample, is the probability that $r_n = p \pm \epsilon$, given that p[S, R] $= p$—symbolically p[$r_n = p \pm \epsilon$, p[S, R] $= p$].

These underlying facts have generated a wide variety of theorems and procedures. For example, Bernoulli proved that the probability of a successful estimate, in this sense, approaches a limit of 1, for any ϵ, however small, as n—the size of the sample—tends to infinity.[1] But it is quite outside the scope of the present book to explore the structure of

[1] The proof in Bernoulli's *Ars Conjectandi* (1713), part IV, ch. v, is expounded in William Kneale, *Probability and Induction* (1949), p. 136 ff.

statistical estimation in any degree of depth or detail. There are in any case plenty of text-books that do this, and the subject itself is mainly mathematical rather than logical or philosophical in character.

It is worth pointing out, however, that, though so-called enumerative induction has at last achieved intellectual respectability by its absorption into statistical estimation, the corresponding function,—viz. the probability of a successful, or unsuccessful, estimate, as described above—is normally employed in some more sophisticated way than as a measure of the credibility of a certain conclusion on certain evidence. It is normally just one element in a procedure for accepting or rejecting hypotheses which is also determined in part by considerations that are quite external to the evidence. So these modern developments can also be regarded as quite in keeping with the older philosophers' contempt for attempts to justify universal propositions solely by the method of enumerative induction.

Two very simple, and convergent, illustrations of this point will suffice: more complicated ones would cloud the issue. The two illustrations will also serve to show that the underlying structure of thought is the same whether we regard so-called enumerative induction primarily as a heuristic procedure for the estimation of magnitudes or as a system of criteria for the acceptance or rejection of hypotheses. Moreover, the illustrations are deliberately taken from two schools of statistical thought between which there has been a good deal of controversy, in order to emphasise that the philosophical point at issue is not affected by these rivalries.

Consider, first, Neyman's method of confidence intervals for estimating magnitudes, whereby the interval of approximation that an estimate is to attain is designated, for a sample of given size, by the probability, which he called the coefficient of confidence,[1] that an estimate based on that size of sample will fall within the interval. Here, in giving an estimate of appropriate approximation from a sample of a certain size, we do not say that the sample provides, say, 95% support for the estimate. Instead, the estimate is said to be correct within 95% confidence limits, and whether we accept the estimate or not depends on whether we regard 95% as a sufficient figure for the purpose in hand. This issue may in practice be determined by purely economic considerations, as in a manufacturer's control of the quality of his products. Suppose a production-run of screws is considered defective only when consumers' threshold of tolerance is exceeded, and suppose that this occurs only when

[1] J. Neyman, 'Outline of a Theory of Statistical Estimation Based on the Classical Theory of Probability', *Philosophical Transactions of the Royal Society of London*, Series A, vol. 236 (1937), p. 333 ff., esp. p. 348.

the screws' standard deviation from their mean length differs more than some particular amount from the advertised standard deviation. Then, in view of his being unable to afford to market more than, say, 5% defective production-runs, the manufacturer may be economically prudent so to adjust the size of the production-batches that he can treat any one of them as a sample which will allow him 95% confidence limits if he estimates from it the standard deviation for the whole of its production-run as precisely as is necessary. Other things being equal, the smaller the percentage of defective production-runs he can afford, the larger must be his coefficient of confidence for estimating the quality of each production-run from the quality of one of its batches.

Now compare Fisher's method of significance levels,[1] whereby we can seek to avoid both the error of rejecting the null hypothesis—the hypothesis that the evidential sample is non-representative—when we should accept it, and also the error of preferring the null hypothesis when we should reject it. For example, if a thing's being an R had no bearing at all on whether it was an S or not, so that p[S, R] = ·5, there would be only a low probability that very much more than half of a sample of a certain size would be composed of R's that were in fact S's: the making of an unsuccessful estimate based on a sample of this nature would then be a relatively improbable event. Hence, if the evidence in fact presented us with a sample composed in this improbable way, we should be more entitled to regard it as significant. By determining in advance the level of improbability at which we shall regard an evidential sample of a certain size as significant we can in effect characterise the type of composition that, if it occurs in our evidence, should make us reject the null hypothesis. If the manufacturer described in the preceding paragraph were a follower of Fisher rather than of Neyman, he would, in effect, need to choose, on the basis of his market research and required expectation of profit, the appropriate significance-level for rejecting a whole production-run on the evidence of a single defective batch. And, if on the Neyman procedure he should require a sample to show the correct standard deviation within 95% confidence limits, or otherwise reject the whole run, on the Fisher procedure he should adopt a significance level of ·05: only samples that have a probability of ·05, or less, on the null hypothesis should make him reject that hypothesis. I.e., if a sample does not show the advertised standard deviation within 95% confidence limits, though it is of the correct size to do so, it must, on the (null) hypothesis that the advertised standard deviation is real, be one of the 5% of samples that fall outside the limits; so that a Fisher-type significance level of ·05 would

[1] Cf. R. A. Fisher, *Statistical Methods for Research Workers*, 7th ed. (1938), p. 120 ff.

lead to the same decision—rejection of the whole run—as a Neyman-type estimate of the real standard deviation that was based on this sample and invoked 95% confidence limits.

It is plain therefore that size of evidential sample is only one of the factors that determine the statistical acceptability of an estimate of an unknown magnitude. Two other important factors are the degree of precision or approximation which the estimate is required to have, and the probability of success it is required to have. These two factors can determine the size of evidential sample that is requisite for an acceptable estimate of a certain kind to be made. But clearly they do not themselves depend upon the observed sample. Instead, so far as they are open at all to rational calculation, they are subject-matter for strategies that will maximise the expectation of gain in one way or another. Now, somewhat analogously, we have already (§10) had to distinguish between evidential support, on the one hand, and acceptability, on the other, in the case of induction by experimental variation of circumstance. But in the latter case the type of supporting evidence to be sought must always be determined by our past experience of relevant variables, whereas in many of the statistical techniques that embrace what used to be called enumerative induction the type of evidential sample to be sought is governed by what we want rather than by what we know.

To sum up so far, then, we are proposing to treat this so-called enumerative induction as a pattern of reasoning that differs from induction by variation of circumstances in at least three ways not recognised by the older philosophers. First, the former pattern is concerned with estimates of probabilities, the latter with support for generalisations. Secondly, the measure of the former is itself a logical probability, while the latter cannot be measured either by a logical probability or by any function of logical probabilities. Thirdly, the former is dependent on non-empirical considerations in a way that the latter is not. Some consequences of these differences are worth tracing out.

First, Hempel's paradoxes disappear here in a different way from that discussed in §11. In the case of inductive support from experimental evidence what was important was to see that though all mutually equivalent propositions have the same grade of support they may not all be testable. I.e. support under any particular support-function is invariant when equivalents are substituted for equivalents; but testability under manipulation of relevant variables is not. In the case of enumerative induction, however, the equivalence principle must be applied with even greater care. The probability functor is not value-invariant under contraposition of its first and second argument-places. Though 'Anything, if it is R, is S' is equivalent to 'Anything, if it is not

S, is not R', $p[S, R]$ is not necessarily equal to $p[\bar{R}, \bar{S}]$ unless $p[S, R] = 1$.[1] Of course, equivalents are substitutable for one another within the probability-functor, e.g. 'R_2' for 'R_1', or 'S_2' for 'S_1', in '$p[S_1, R_1]$'. But we are not normally entitled to make the substitutions that would transform '$p[S, R] = p \pm \epsilon$' into '$p[\bar{R}, \bar{S}] = p \pm \epsilon$'. So there can be no paradox about the fact that evidential samples which have a high probability if $p[S, R] = p \pm \epsilon$ may have only a low probability if $p[\bar{R}, \bar{S}] = p \pm \epsilon$. Again, we must always have $p[R \ \& \ \bar{R}, R \ \& \ \bar{S}] = 0$, even when $p[S, R] = 1$. So no evidential sample can ever provide a basis for estimating that the value of $p[R \ \& \ \bar{R}, R \ \& \ \bar{S}]$ is close to one even when such a basis is available for $p[S, R]$. Indeed it is worth emphasizing that the probabilistic counterpart of $(x)((Rx \ \& -Sx) \rightarrow (Rx \ \& -Rx))$ must always have zero value even when the probabilistic counterpart of $(x)(Rx \rightarrow Sx)$ has its maximum value. For this shows that we are barred from conceiving all generalisations to be hypotheses of maximum probability, as some philosophers have proposed.

Secondly, Goodman's paradoxes also disappear here in a different way from that discussed in §11, and in a way that is also essentially dependent on the assumption that enumerative induction is not concerned to support generalisations but to estimate probabilities. In the case of inductive support from experimental evidence the danger of allowing positive support for both of two incompatible predictions—that all emeralds examined after moment *m* are blue and that all emeralds examined after *m* are green—is averted by insisting on the primary relevance of the variable directly involved, viz. (in Goodman's example) the time variable. No clear positive support is to be allowed unless it stems from a test in which at least the most relevant variable is manipulated. But insofar as

[1] It is demonstrable that in a finite universe containing u things in all, where $p[S, R] < 1$ and r is the number of R's and s the number of S's, $p[S, R] = p[\bar{R}, \bar{S}]$ if, and only if, $u = r + s$. Let x be the number of R & S's (i.e., of things that are both R's and S's), y the number of $\bar{R}$ & $\bar{S}$'s (i.e., of things that are neither R's nor S's) and z the number of R & $\bar{S}$'s (i.e., of things that are R's but not S's). We first prove that, if $u = r + s$, then $p[S, R] = p[\bar{R}, \bar{S}]$. By elementary set theory $x = r - z$, and $y = u - s - z$. Hence, if $u = r + s$, $x = y$. Also, if $p[S, R] < 1$, then $z > 0$; so both $r > 0$ and $(u - s) > 0$. Therefore, if $x = y$ and $u = r + s, \frac{x}{r} = \frac{y}{u - s}$. Hence, if $u = r + s$, $p[S, R] = p[\bar{R}, \bar{S}]$. Secondly, we prove the converse of this. If $1 > p[S, R] = p[\bar{R}, \bar{S}]$, then $\frac{x}{r} = \frac{y}{u - s}$. But, by elementary set theory again, if $\frac{x}{r} = \frac{y}{u - s}, \frac{r - z}{r} = \frac{u - s - z}{u - s}$. Hence, if $p[S, R] = p[\bar{R}, \bar{S}]$, $u = r + s$. So a finite universe must be very nicely adjusted in size if straightforward contraposition is to be made admissible for the dyadic probability-functor. Cf. also H. Reichenbach, *The Theory of Probability* (1949), p. 435, and P. Suppes, 'A Bayesian Approach to the Paradoxes of Confirmation', in *Aspects of Inductive Logic*, ed. J. Hintikka and P. Suppes (1966), p. 205 ff.

the measure of enumerative induction—the probability of a successful estimate of a probability—is related to the mere size of an evidential sample, and not to experimentally varied circumstances, this method of avoiding paradox is not available in relation to enumerative induction. Instead, we avoid the sensation of paradox here by noting that the probabilities we estimate do not license predictions in the way that generalisations do. Even on the basis of a statement that $p[S_1, R_1]$ is within a very small interval indeed of 1, we still cannot predict anything about a specific object that happens to be an R_1. We cannot say whether the object will, or will not, so far as we can foresee, be an S_1. For we have to bear in mind that it may also have another characteristic R_2, such that $p[S_2, R_2]$ also has a high value and being an S_2 is incompatible with being an S_1. The very same object about which we are tempted to predict, because it is emerald, that it will be grue and therefore blue after *m*, may also be a member of the class of green or red objects that do not change their colour at *m*; and members of that class have quite a different probability of being blue after *m*. This is a familiar feature of probability-statements and we can handle it by determining which features of objects are relevant, at a given level of significance, for statistical hypotheses generating the type of credible prediction we wish to make (cf. p. 118 ff.), and then estimating the probability of the predicted characteristic for the subset of these features that the object in question possesses. On such a basis, e.g., an appropriate life insurance premium can be estimated for a man who is not only aged 30 but also a lorry-driver and a diabetic.[1] But to do all this here requires very much more evidence than just the pre-*m* sample of emeralds that are both green and grue. So we need have no fear that so-called enumerative induc-

[1] It is sometimes said that in such cases we need to determine the probability of the event to be predicted 'on the total available evidence'. But if this means 'on the basis of everything that anyone knows' it is quite unrealistic. No-one can cope with such a mass of data. If it means 'on the basis of a sample from the class of all those things which share the same conjunction of characteristics as the object in question' it is equally unrealistic, since even if we except spatio-temporal location the object may well be unique. And, if it means 'on the basis of a sample from the class of all those things which share the same conjunction of *relevant* characteristics', it is less misleading to say just this. C. G. Hempel's otherwise lucid discussion of this issue in 'Inductive Inconsistencies', *Synthese* xii (1960), p. 439 ff. (reprinted in his *Aspects of Scientific Explanation*, 1965, p. 53 ff.), is impaired by its conception of all inductive reasoning as being enumerative or probabilistic. In its enumerative or probabilistic form the problem of so-called 'inductive inconsistencies' is to be solved, as emerges in the present chapter, by recourse to experimental, variative induction. We have to choose our samples from those with the same inductively relevant characteristics. But the problem also arises, as was shown in §8, in relation to experimental, variative induction; and there its solution depends on the proper analysis of reports of canonical test-results and on the recognition that all assessments of inductive support are empirically corrigible. If two true evidential reports support inconsistent judgements, a hidden variable must be operating and our criteria of assessment require revision (cf. §§8–9).

tion can justify incompatible predictions on the basis of such a sample, and no reason therefore to seek grounds for banning terms like 'grue' from the vocabulary that we use to formulate estimates of probabilities.

Thirdly, if enumerative induction is indeed a method of estimating probabilities in which size of evidential sample is the sole criterion of a sample-based estimate's merit, it cannot license the derivation of what it is convenient to call ampliative counterfactuals. An ampliative counterfactual implies that such-or-such would be the case even if more things satisfied a certain condition than in fact do, have done, or will do. Familiar examples stem from causal laws, like 'If you had walked on the ice you would have broken it'. Ampliative counterfactuals are thus to be sharply contrasted with the counterfactuals stemming from accidental truths. For example, even if it is only accidentally true that all three of Smith's dogs are white, we are still entitled to say, when someone sees a brown dog and suggests that it is Smith's: 'If that had been one of Smith's [three] dogs, it would have been white'. But because it stems from an accidental truth this has to be a non-ampliative counterfactual: it does not tell us what colour the dog would have been if it had been Smith's fourth dog.

Now, ampliative counterfactuals are rather like predictions, in respect of their derivability from probability-estimates. Even if p[S, R] is estimated, on the basis of a large sample, to be within a very small interval of 1 we can infer nothing from this about things that are not R's. The estimate is just about the actual population of R's. So we are not entitled to say what the probability is, that some specified thing, which is actually not an R, would have been an S, on the sole assumption that it had in fact been an R. Before we can say anything here we need to know what other characteristics the specified thing is supposed to have; which of these characteristics are relevant in relation to being an S; and what is the estimate of the probability that R's having all these characteristics are S's. When we are also told his occupation and his state of health, for example, we may be able to calculate what *would have been* an appropriate life insurance premium for a man if he *had been* 30 years old. Hence no enumerative induction on its own, however large its evidential sample, can license derivation of an ampliative counterfactual. Indeed the procedure of invoking relevant characteristics here is syntactically analogous, as will be shown later (§16, esp. p. 152 f.), to the manipulation of relevant variables. In the latter case we test a broadly formulated hypothesis for its validity in inductively relevant circumstances V_1, V_2, etc.: in the former we restrict our hypothesis to certain inductively relevant circumstances V_1, V_2, etc. for which it is found to be valid. But we certainly cannot obtain empirical backing for hypotheses that generate ampliative counterfactuals without appealing at some point to the inductive relevance of certain variables.

Thus the hypotheses from which ampliative counterfactuals are characteristically derivable are generalisations that enjoy inductive support. In fact we can characterise in just this way the crucial difference between a generalisation that we take to be accidentally true, if true at all, and a generalisation that we take to be true as a law of nature (or a consequence of one), if true at all. Philosophers have sometimes suggested that science has no business to concern itself with generalisations referring to particular, specified regions of space or time. But this puts the point wrongly. There is no reason at all, as was remarked in regard to 'grue', why scientific generalisations should not refer to the planets of our own solar system, say, or the plant-life of the mesozoic period. No doubt even more support for such generalisations may be established when they have been unified or incorporated into theories of much greater generality—theories that do not refer to particular, specified regions of space or time. But what is crucial for the status of generalisations as established or putative laws (or consequences of laws) is not the nature of the terms in which they are formulated: it is rather the nature of the evidence that exists or is expected for them. This evidence must be achieved by the manipulation of relevant variables, and the consequential elimination of rival hypotheses, rather than by what the older philosophers called enumerative induction.[1] Mere numbers of actual R's, irrespective of their circumstances, cannot suffice to support any thesis about what other things would have been like if they too had been

[1] Hence to regard a generalisation as being open to inductive tests is to regard it as a potential law. Elsewhere I have referred to the feature of a generalisation that makes it a potential law as 'precisifiability': cf. *The Diversity of Meaning*, 2nd ed. (1966), ch. X. This is not inconsistent with what is being said here, but from the standpoint of the present book precisification would be a non-standard form of inductive reasoning. Compare what is said about simplification-functions in §16 below.

Though there are some forms of generalisation that leave it open to us whether to consider them as potential laws or only as accidental truths, there are others (where 'any', e.g., occurs instead of 'every' or 'must be' instead of 'is') that pre-empt the issue. Similarly there are some forms of counterfactual conditional that leave it open to us whether to consider them as ampliative or non-ampliative (e.g. 'If Smith had been one of the heart transplant cases, he would not have recovered') and others which pre-empt the issue (e.g. 'If Smith had been one of the heart transplant cases, he could not have recovered'). If a sentence is asserted informally with the intention of advancing an inductive hypothesis it is normally formulated unambiguously as a potential law (e.g. as in §§5–9: cf. p. 7, n. 1). But to offer an inductive hypothesis is tantamount to saying that a certain proposition—which does not claim law-like status for itself—is to be treated as one for which inductive support may in fact exist. So in the formal syntax of induction I shall represent such a proposition by a truth-functional conditional, and my object will be to deploy what implies or is implied by the existence of support for such a proposition (cf. §§20–22 below).

R's. There is always the risk that the observed R's were all specially circumstanced in some relevant way in which the other things were not.

§13. Support-grading for statistical hypotheses

Many statistical hypotheses, though of an apparently non-generalising character, are capable of acquiring inductive support *via* experimental tests on equivalent generalisations of a certain kind. But this support is to be assessed, as in the case of non-statistical hypotheses, solely by reference to test-results. Statistical criteria, such as significance-levels, serve to indicate what is acceptable as inductive evidence, rather than to measure inductive support. Indeed, many support-functions must be taken to be defined over some such criterion as well as over a particular class of materially similar hypotheses. But their logical syntax is quite unaffected by this. Nor is it possible to devise a measure of credibility that would somehow amalgamate the probability estimated *by* a hypothesis with the grade of inductive support that exists *for* the hypothesis.

We have so far discussed statistical rules for estimating magnitudes, or for accepting or rejecting hypotheses, in abstraction from the establishment of inductive support by experimental tests. But, though the underlying structures of both estimation and induction are revealed more clearly in that way, the two modes of reasoning are often so closely intertwined in practice that it is also necessary to lay bare some of the patterns in which this intertwining occurs. Otherwise the reader may still feel tempted, in certain complex situations, to suppose that the measure of inductive—i.e. experimental—support is a probability or function of probabilities.

What made it possible to discuss inductive support in abstraction from statistical estimation was the assumption that elementary inductive hypotheses are formulated in terms which enable experimenters to observe whether each individual object or event, as it presents itself, satisfies both the antecedent and the consequent of the hypothesised generalisation. In particular, this was the assumption in terms of which the ranking or grading of inductive support was discussed (in §§5–8). But even the most elementary hypotheses in modern science are often formulated in terms of statistical magnitudes, such as probabilities, means, medians, variances, etc. For example, we may wish to hypothesise, quite precisely, and with 95% confidence, that for a 30-year old healthy male European the probability of death before 70 is $\cdot 68 \pm \cdot 01$. Or instead of just claiming that a certain drug is a good treatment for such-or-such an infection, we may prefer to claim that patients to whom appropriate doses of it are administered have a significantly higher probability of recovery (with a specified significance level) than patients to whom no treatment is given. Nor are references to statistical magnitudes found only in relatively elementary hypotheses. They are common too in what were called above (§9) correlational generalisations. For example, a pharmacologist may wish to hypothesise, with 95% confidence limits, that the median lethal dose of a certain drug for any species of mammal is such-or-such a function of that species' mean weight. Indeed, wherever a measurable quantity is at issue, experimenters may prefer to be interpreted as speaking of the mean, say, or the median in a population of measurements rather than of a single uniform measurement: they can thus allow for apparently unimportant fluctuations that are not readily amenable to control.

At first sight it might seem that statistical hypotheses such as these fall quite outside the scope of inductive tests, as the latter have so far been envisaged. The testable generalisations discussed in §§5–9, whether elementary, causal, correlational or theoretical, were all universally quantified conditionals. But most statistical hypotheses seem essentially different from those generalisations in that they seem to ascribe a certain magnitude to a population as a whole rather than to say something about each member of it.

However, there is no real difficulty here. Any statistical hypothesis has at least one equivalent that is a universally quantified conditional. For instance, any statistical probability-statement of the form

(1) $\mathrm{p}[\mathrm{S}, \mathrm{R}] = p \pm \epsilon$

has an equivalent generalisation of the form

(2) $(\mathrm{V})((\mathrm{V} \subset \mathrm{R}) \rightarrow (\exists \mathrm{W})((\mathrm{V} \subset \mathrm{W}) \;\&\; (\mathrm{W} = \mathrm{R}) \;\&\; \mathrm{p}[\mathrm{S}, \mathrm{W}] = p \pm \epsilon))$

where (2) is to be read 'For any set V, if V is included in R, there is a set W such that V is included in W and W is co-extensive with R and $\mathrm{p}[\mathrm{S}, \mathrm{W}] = p \pm \epsilon$'. Admittedly, generalisations like (2) may, as we have seen, be merely accidental truths. For instance, if 'S' denotes the set of white things and 'R' the set of Smith's dogs, and if $p = 1$ and $\epsilon = 0$, both (1) and (2) are equivalent to 'All Smith's dogs are white'. But the hypotheses we must concern ourselves with here are not like this. If true, they are not accidentally true. They can act as premisses for the deduction of ampliative counterfactuals. From the hypothesis that patients treated with such-or-such a drug have a significantly higher probability of recovery from a certain infection than patients not so treated, it is deducible that if Dr. A's patients had been so treated they too would have belonged to the class of patients with a significantly higher probability of recovery. In other words, many probability-statements like (1) must be capable of enjoying inductive support from the results of tests in relation to relevant variables. It follows that many statistical hypotheses, if not testable themselves, must have inductively testable equivalents, like (2). Such hypotheses are just as much capable of acquiring inductive support, through successful tests on equivalents, as are non-statistical ones.

It is easy to see how a generalisation like (2) can be tested in ways closely analogous to those described in §§5–9 for non-statistical hypotheses. In the latter case a successful trial-outcome of a test t_i, where $i \geqslant 1$, requires that some member of the hypothesised generalisation's domain of discourse should satisfy both its antecedent and its consequent in a particular combination of variants of relevant variables, and should be unaffected by other factors than variants of the variables being manipulated. Correspondingly in the case of (2) what is required is that there should be some set of objects or events that satisfies three conditions: it must be a sample of R on which an estimate $\mathrm{p}[\mathrm{S},\mathrm{R}] = p \pm \epsilon$ may be based, it must be selected under some particular manipulation of relevant variables, and its composition must be unbiased by other factors than variants of the variables being manipulated.[1] The size of this sample must be determined by the content of the estimate together with some such *a priori* stipulation as a coefficient of confidence. We cannot, of course, find any analogies for this *a priori* stipulation if we look at the testing of non-statistical generalisations, as described in §§5–9. But a successful trial-outcome requires us to accept that both the antecedent and the consequent of the hypothesised generalisation have been satisfied, and where the consequent asserts an estimated or estimatable magnitude we obviously need to invoke some suitably rewarding criterion of acceptability in order to tell when a trial has had a successful outcome.

Such a criterion must not be conceived of as a measure of the evidential support for the hypothesis as a whole, but rather as an indication of the nature of the trials from which this-or-that grade of support is sought or attained. To have a higher coefficient of confidence for the same estimate we need a larger sample, and this may cost time, trouble and money. But normally, if we are prepared to pay the price, a larger sample is in fact obtainable and a higher confidence-criterion can be satisfied.

[1] Such variants may consist not only in each of the sample's members' having a certain characteristic, but also in a certain proportion of these members' having one characteristic, a certain proportion another characteristic, and so on, as in a so-called 'structured sample'. (If a hypothesis fails a particular test t_i, a more heavily qualified version of it may be more successful: cf. §16 below.)

A higher grade of support, on the other hand, may not be so readily available. Whether we can achieve it or not depends on the operation of the relevant natural variables as well as on the price we ourselves are prepared to pay for a performance of the appropriate test. The grade of inductive support attained by a hypothesis is determined solely by the extent to which it remains unfalsified under the manipulation of a cumulative hierarchy of relevant variables. In short, if U is like (2), and E describes the outcome of trials on U, the dyadic support-function s[U, E] cannot have a positive value if one of the samples described in E is not of a size appropriate to the estimate in U and the stipulated confidence-level: E does not then describe a canonical test on U. But if these samples are all of an appropriate size, then s[U, E] has its value determined, in the normal way, by the trial outcomes reported in E.

For example, in testing whether, for a 30-year old healthy male European, the probability of death before 70 is $\cdot 68 \pm \cdot 01$, the most relevant variable for that particular class of materially similar hypotheses may be thought to be the medical history of the person's parents. If appropriately chosen samples, e.g. from people with healthy parents and from people without, still generate an estimate within this interval, at the stipulated confidence-level, the hypothesis has passed the second test and acquired second-grade support. But if there was only one, randomly selected sample, the hypothesis could have passed only the first test and acquired only first-grade support. Of course, in practice experiments of such a type are very often designed with a heuristic purpose. They are very often designed to provide a basis for making a well-supported estimate rather than to test an already conjectured one. But the support-relation between a conjunction of evidential propositions, on the one side, and a testable hypothesis, on the other, is quite independent of any contextual assumptions. It does not depend on the purpose with which the evidence has been ascertained or on whether we entertain the hypothesis before we observe the evidence (cf. §1 above).

Also, it is always arguable in relation to a hypothesis like this, that the probabilistic character of the hypothesis betokens our ignorance of many of the relevant variables. If only we knew all the factors affecting date of death, for example, it might be said that we could generalise deterministically about the life span of each type of person in each type of situation. But since complete knowledge of relevant variables in such a case would require not only a complete human genetics and epidemiology, but also a complete meteorology, a complete psychology and sociology of human behaviour, and much else besides, the possibility is quite an unreal one. So we cannot, at least in practice, avoid the need to seek inductive support for statistical hypotheses, like (2), that generalise over classes, not individuals. However, it is interesting to note the way in which the evidential '$T_i x$'—cf. (2) of §7—may be substantiated in the case of such hypotheses. Since the elements of (2)'s domain of discourse are classes, not individuals, the evidential instances will also be classes. So we can use random selection of the members of these evidential samples as a way of trying to ensure that the characteristics of the evidential instances are not affected by any circumstance that is not either a circumstance mentioned in the antecedent of the hypothesis undergoing test or a variant of the variables manipulated in the actual test performed.

A somewhat more complicated procedure would be needed in order to build up inductive support for a hypothesis that treatment with a certain drug gives patients a significantly higher probability of recovery from such-or-such an infection than does the absence of any treatment at all. The analogy now is not with the quite elementary generalisations discussed in §§5–8, like 'Anything, if it is R, is S', but with the causal generalisations discussed in §9, like 'Being R is a cause of a Q thing's becoming S'. The difference, from the latter, is just this: instead of saying that a certain drug always cures people of such-or-such an infection, the hypothesis now under consideration says just that it gives them a significantly higher probability of recovery. Accordingly, as in the case of causal generalisations, we need a control for each trial that is to constitute part of a canonical test. Each control sample provides a basis for estimating the probability of spontaneous recovery in particular combinations of variants of relevant variables, such as age, previous infection, etc. We may regard the null hypothesis for each trial as stating that this control-based estimate represents the real probability of recovery in any case. The corresponding trial will then have a successful outcome if it operates on a sample of treated patients from which a significantly different probability of recovery may be estimated, i.e. if the sample has a significantly low probability of being selected on the assumption that the null hypothesis is correct. We need to be sure, for a trial-outcome to be successful, that the factors operative in producing recovery lie in the treatment given rather than in the particular combination of relevant circumstances characterising the sample. In other words, we need to use something like Mill's method of difference here as well as something like his method of agreement, as in the case of non-statistical causal generalisations. But we have in addition to invoke some *a priori* criterion of significance, which will in effect determine—in the light of information supplied by the corresponding control—whether a particular trial-outcome is to be regarded as successful or not. So the significance-criterion is not a measure of evidential support, but rather an indication of what would be acceptable as supporting evidence. A stricter significance-criterion (along with a reduced probability of accepting the null hypothesis when in fact it is false) can be bought fairly cheaply, by increasing the size of the samples investigated. But this does nothing to raise the grade of inductive support for the hypothesis under examination. Whether or not the support-grade can be raised depends at least as much on how pharmacological variables operate as on the time and patience of pharmacologists. Increases in inductive support are not just a matter for human decision.

It would be pointless to consider here more complicated examples than these. No fresh philosophical problems, for instance, seem to be raised by the incorporation of statistical hypotheses into correlational generalisations. The present book is not offered as a text-book of experimental method, but as an analysis of the logical syntax of inductive support. The purpose of investigating some of the patterns in which statistics gets intertwined with inductive support-assessment has been to show that the fundamental structure of the latter is not affected by its association with statistical procedures. Its logical syntax is the same here as elsewhere. The general conjunction principle, for example, that was established in §8 is just as valid when the support-functor has its hypothesis argument-place filled by a statistical generalisation like (2) as when the support-functor deals with a non-statistical generalisation, since both types of generalisation acquire inductive support in fundamentally the same way—by escaping falsification under the manipulation of cumulatively increasing combinations of relevant variables.

It may perhaps be thought that quantification over sets in statistical generalisations like (2) excludes support for them from the scope of an instantial comparability principle, since quantification over kinds, characteristics or circumstances certainly excludes scientific theories and correlational generalisations from the scope of such a principle, as was shown in §9. But in the latter case two different correlational generalisations U and U′, could have one or more substitution-instances in common, even if U was much better supported than U′ by test-results, whereas if two statistical generalisations like (2) share a substitution-instance they must also be logically equivalent to one another and thus have the same support as one another. So there is no analogue here, in the case of (2), for the argument (see §9) that prevents any instantial comparability principle from governing support for correlational generalisations or scientific theories; and everything that can be said in favour of the instantial comparability principle in the case of elementary non-statistical generalisations can also be said in favour of it in the case of elementary statistical generalisations. Correspondingly we must also accept the instantial conjunction, and uniformity, principles in the latter case.

Two other points need to be made here. The first is that we have to amplify our account of how a natural variable is shown to be inductively relevant to a particular class of materially similar hypotheses. What was said before (in §5) was that one or more of those hypotheses must both have been falsified in some situation when one variant of any two-variant variable in question was seen to be present, and also have been positively instantiated when the other variant was present. But it is clear now that falsification of hypotheses here must be taken to include rejection of statistical generalisations in the light of *a priori* stipulations about confidence-intervals, significance levels, etc. It follows that many support-functions must be taken to be defined over some such *a priori* stipulation as well as over a particular class of materially similar hypotheses. The former element in the definiens, as well as the latter, can affect which variables are relevant and, therewith, which tests are appropriate. Moreover, a support-function thus defined may be conceived of as determining the requisite confidence-interval, significance-level, etc. for canonical tests that are appropriate to it. The use of any one such support-function imposes a constraint not only on which hypotheses may be said to have positive support but also on which evidential samples may be said to afford successful trial-outcomes under the manipulation of relevant variables.[1]

Secondly, if we establish full inductive support, with suitably chosen statistical criteria, for the hypothesis that treatment with such-or-such a drug has a high probability of success, we have certainly made it more credible that any particular patient so treated will be cured than if we had only established full support for a lower probability of success. Also, if we establish jth-grade support for a certain probability of success we have certainly made it more credible that any particular patient so treated will be cured than if we had established only ith grade support, where $j > i$, for the same probability of success. But it does not seem possible to devise a rationally defensible measure of credibility here that would, as it were, amalgamate the probability with the grade of support. On some occasions the two credibilities to be compared might stem, in

[1] It also imposes an implicit restriction on the extent to which experimental results may be assumed to be repeatable, or on the statement of what is said to be repeatable. If a sample of a certain size has only a ·95 probability of being representative, within a specified interval, of a certain magnitude in a population, then experimental results based on that size of sample, randomly selected, can be ascribed a corresponding probability of occurrence within any run of attempted repetitions; or what is repeatable can be said to be a run of experimental results within which one kind of result has that probability of occurrence.

the one case, from high-grade support for a low probability, and, in the other case, from low-grade suport, according to the same support-function, for a high probability. An appropriate credibility-function would have to make any two such cases commensurable, and also take into account differences in *a priori* stipulations about confidence-intervals, significance-levels, etc. But there seem to be no plausible criteria for evaluating the comparative appropriateness of two different credibility-functions for this kind of task, where one of the two attaches more weight to grade of support and the other to probability (along with appropriate stipulations about confidence-intervals, significance-levels, etc.). Statistical criteria have already been invoked in interpreting the evidence for inductive support; the criterion of exposure to future experience bears on the tenability of the support-function; and an appeal to intuition is just as unsatisfactory here as it is for judging measures of inductive support (cf. §6, p. 50 f., above). It looks therefore as though we must do without a general credibility-function of this kind, and be content with just comparing different probabilities that have the same positive grade of support, or different grades of support for the same probability.

§14. A historical excursus

> If the analysis proposed in this book is to be placed in its proper historical perspective it should be regarded as an attempt to reach a higher level of philosophical sophistication within the tradition of inductive logic that began with Bacon's *Novum Organum* and was continued by Whewell, Herschel and Mill. Hume's preoccupation with the causal processes analogous to enumerative induction was the archetype both for the obsession with stimulus-response learning that has obstructed psycho-linguistic investigations and also for the obsession with the probability-calculus that has issued in philosophical theories of induction like those of Keynes, Ramsey, Nicod and Carnap. The betting analogy sometimes invoked by such philosophers, as a criterion of adequacy, is quite inapplicable to the central core of scientific reasoning. Experimental support is not to be viewed as a species of a logical genus to which enumerative induction also belongs, but rather as a species of a quite different genus syntactically definable, which requires separate investigation.

This book has no polemical intent. It is primarily concerned to state and defend certain propositions about the logical syntax of support-functions, not to criticise other philosophers' views on the subject. Nevertheless some readers may find it easier to accept the arguments here, if they are seen in rather more of their historical context than has hitherto been depicted.

Francis Bacon, in his *Novum Organum*, was the first philosopher to tabulate the fundamental logical ideas of experimental enquiry. He urged scientists to interrogate Nature, instead of just waiting upon the observations that came their way, with the object of constructing tables of presence and absence—lists of circumstances present, and circumstances absent, when the phenomenon under investigation was found. When a true uniformity had been thus established, it was itself to be treated in turn as just one variety of phenomena to be explained by some deeper and more comprehensive law. But this theme, of induction by variation of circumstance, was curiously ignored by John Locke, in the somewhat opaque discussion of sensory knowledge to be found in his *Essay on Human Understanding*.[1] It is understandable that Locke should have been sceptical about the possibility of deriving universal truths from sensory perceptions by deductive argument. But his own views would have been clearer if he had at least discussed Bacon's claim to be able to do this by inductive argument.

David Hume tried to ride both horses at once. Sometimes he writes in the Baconian tradition of inductive logic. At other times, perhaps only a few pages away, he writes in the Lockeian tradition of scepticism about the possibility of reasoning from the observed to the unobserved. On the one hand, he subtitled his *Treatise of Human Nature* 'An attempt to introduce the experimental method of reasoning into moral subjects', and included a section headed 'Rules by which to judge of causes and effects' in which he claimed to set forth what he called the 'logic' of experimental reasoning. Though his treatment of the subject was very superficial, compared with Bacon's, he obviously recognised the importance of varying circumstances. 'There is no phaenomenon in nature,' he wrote, 'but what is compounded and modify'd by so many different circumstances, that in order to arrive at the decisive point, we must carefully separate whatever is superfluous, and enquire by new experiments, if every particular circumstance of the first experiment was essential to it'. On the other hand, and quite inconsistently, Hume also asserted that no causal uniformity can be an object of reasoning or operate upon the mind in any way 'but by means of custom'.[2] Indeed in his sceptical passages Hume pays no attention at all to the characteristic features of experimental reasoning. It is not just that he prefers to offer

[1] See especially book IV, chapters i–iii. On some sketchy medieval antecedents of Bacon's work, in the writings of Robert Grosseteste, Albertus Magnus, Duns Scotus and William of Ockham, cf. J. R. Weinberg, *Abstraction, Relation and Induction* (1965), p. 121 ff.

[2] Cf. *Treatise*, Bk. I, pt. III, sec. xv, Selby-Bigge edition (1888), p. 175 and Bk. I, pt. III, sec. xiv, Selby-Bigge ed. p. 170. Hume's scepticism about induction is discussed further in §19 below.

here a causal explanation of our general beliefs about matters of fact, rather than a logic for them. He is also curiously narrow even in the type of cause that he is prepared to recognise. He talks about the effect on a man's mind of constantly observing one kind of event to succeed another, not about the effect of observing one kind of event to succeed another in a variety of different circumstances. He seems obsessed with the kind of causal influencing of the mind that is analogous to what logicians like Bacon called induction by simple enumeration, and ignores altogether the kind of causal influencing that is analogous to induction by variation of circumstances.

Hume's preoccupation with the causal analogue of enumerative induction is the direct ancestor of a similar tendency in certain twentieth-century psychologists' theories of learning. To explain learned behaviour by reference to a repeated pattern of stimulation and rewarded response, for example, is to confine one's explanatory model to processes of enumerative induction. Not surprisingly that model has been found quite inadequate to explain certain distinctively human types of learning, such as the acquisition of an ability to speak and understand one's native language. In particular, it seems impossible to explain thus how children learn to speak grammatically. If they do learn to do this by processes analogous to enumerative induction their brains must presumably be supposed to register very many of the probabilities with which a string of words belonging to certain categories is followed by a word belonging to a certain category. But it has been shown[1] that the short span of human childhood is nowhere near long enough for such a Markovian learning-process to take place. Enumerative induction is not a suitable model for language-learning.

Nevertheless, those who on this account rightly criticise any psychologists who adhere to that model are themselves wrong to suppose that the only alternative is to reject altogether the view that fundamental laws of grammatical structure are learned from experience. To suppose this is to continue Hume's preoccupation with the causal analogue of enumerative induction. If so-called enumerative induction is indeed the only way to learn from experience, then admittedly considerations of time suffice to show that the fundamental principles of grammatical structure must be supposed innate. But if children learn the grammar of their native language through a tacit process of induction by variation of circumstance, there is no need on account of temporal considerations

[1] N. Chomsky and G. Miller, 'Finitary Models of Language Users', in *Handbook of Mathematical Psychology*, ed. R. D. Luce, R. Bush, and E. Galanter, vol. II (1963), p. 424 ff. Cf. also N. Chomsky's review of B. F. Skinner, *Verbal Behaviour* (1957), in *Language* XXXV (1959), p. 26 ff.

to postulate any innate principles that are solely and specifically concerned with grammatical syntax. The formation and testing of rules, or universal hypotheses, is a much more time-saving procedure, in at least three respects, than the piecemeal collection of Markovian probabilities.[1] First, a single speech-experience, when a child hears a corrected version of its own utterance, is only a drop in the ocean of evidence that the latter procedure requires, but the former type of procedure allows even one such speech-experience to eliminate a whole class of possible structures. The force of a negative instance, as Bacon saw, can be very great. Secondly, the more heavily we qualify our universal hypotheses by mentioning some combination of variants of relevant variables in their antecedents, the fewer the trials that we need in order to find out whether or not those hypotheses are fully supported, as emerges in §16 (p. 150) below; and most elementary grammatical hypotheses, describing the shared syntax of a class of grammatical sentences, need to be fairly heavily qualified in such a way. Thirdly, once we have evidence to support a suitably wide variety of elementary hypotheses in a particular field of enquiry, we thereby have evidence to support a theory (if we can think of one) that will not only unify these hypotheses but also enable us to infer many, perhaps infinitely many, other elementary hypotheses (cf. §9, p. 83 ff. above).

Our purpose here, however, is to investigate inductive reasoning as an articulate logical structure that is implicit in the conscious arguments of experimental scientists. We are not primarily concerned with the use of this structure as a postulatable model for the psychological explanation of such behavioural phenomena as childhood language-learning. But twentieth-century philosophers have unfortunately been as much influenced as psychologists by Hume's one-sided legacy.

The first half of the nineteenth century saw some brief progress in the development of Bacon's seminal ideas. William Whewell discerned the importance of what he called consilience in the construction of scientific theories, and his researches in the history of science enabled him to illustrate this from post-Baconian examples. Though he was perhaps more Kantian than empiricist in his philosophy of physics, Whewell's notion of consilience was, as we have already seen (cf. §9 above), a form of induction by variation of circumstances. Moreover, in his respect for the actual historical facts of scientific enquiry, Whewell's mode of philosophising was far more empirical than that of either Locke or Hume.

[1] Cf. L. Jonathan Cohen, 'Some Applications of Inductive Logic to the Theory of Language', forthcoming, in *Amer. Phil. Quart.* vii (Oct., 1970). On the relation of induction to a language-learner's production of semi-grammatical sentences cf. §18 below. On the importance of the concept of a rule in the psychology of learning cf. G. A. Miller, F. Galanter and K. H. Pribram, *Plans and the Structure of Behaviour* (1960), *passim*.

His friend, J. F. W. Herschel, resembled him here. Placing a picture of Bacon's bust on the title page of his *Preliminary Discourse on the Study of Natural Philosophy*, Herschel used a wealth of post-Baconian examples to illustrate the basic soundness of Bacon's ideas. On the one hand he held that if there is a single circumstance in which all the facts without exception agree then that circumstance is certainly the cause in question, or at least a collateral effect of the same cause: on the other hand, he held that the prime purpose of experimentation is to find instances differing in one circumstance and agreeing in every other.[1] A decade or so later J. S. Mill, in his *System of Logic*, sharpened this antithesis between what he called the method of agreement and the method of difference, and investigated some of the problems to which it gives rise.

None of the older writers distinguished clearly and systematically between the concept of induction as a method of discovering acceptable solutions to scientific problems and the concept of induction—or rather, of inducibility—as the converse of a logical, or quasi-logical, relation of support between propositions. So Bacon, for example, suggested that experimenters should normally collect a great deal of evidence before even formulating any hypotheses; and criticism of this rather unfortunate methodological proposal has sometimes distracted attention from Bacon's very important logical, or quasi-logical, insights into the nature of the evidence that is required to support a scientific hypothesis. Indeed, some modern critics of Baconian methodology have attempted to confine the term 'induction' to a methodological sense, and have accordingly coined the term 'inductivism' as the name of Bacon's methodological fallacy. But the analogy with 'deduction', 'deductive', etc. is sufficient to justify our continuing to use the terms 'induction', 'inductive', etc. in a logical or quasi-logical sense.[2]

Since they thought of induction as a method of discovery, where the alternatives were just success or failure in the discovery of truth, the older writers did not usually conceive of inductive support as admitting a hierarchy of grades in accordance with differing degrees of thoroughness

[1] *A Preliminary Discourse on the Study of Natural Philosophy* (1833 ed.), pp. 152 and 155. The method of grading inductive support that was described in §§7–9 above can be regarded as a detailed working out of the consequences of Herschel's remark (ibid. p. 155) that experiments 'become more valuable, and their results clearer, in proportion as they possess this quality (of agreeing exactly in all their circumstances but one), since the question put to nature becomes thereby more pointed, and its answer more decisive'.

[2] J. M. Keynes remarked as long ago as 1921, in his *A Treatise on Probability*, p. 265, that the importance of Bacon and Mill, in relation to induction, lay in their contribution to its logic, not in their views about the methodology of scientific discovery. Mill himself wrote, in the third and later editions of his *System of Logic* (III, ix, 6) that, even if his inductive methods were not methods of discovery, it would not be the less true that they were methods of proof.

of appropriate tests, and they therefore made no attempt to explore the logical syntax of such gradations. Indeed, they did not normally show awareness of the fact that the appropriateness of a test is itself an empirical issue, and they often made exaggerated claims on behalf of their methodology. But apart from the exposure of some of these exaggerations, and the criticism of such non-Baconian theories as Reichenbach's or Carnap's, it is very difficult to identify any substantial contribution that has been made to the further development of Bacon's ideas during the century and a quarter or so since the publication of Mill's *System of Logic* in 1843.

What has happened instead is that, just as Hume was largely preoccupied with enumerative induction, so too modern philosophers of induction have been largely preoccupied with inductive measures that are either probabilities or functions of probabilities. For example, J. M. Keynes in his *Treatise on Probability* (1921) attempted, as it were, to mix oil and water by devising a probabilistic measure for induction by variation of circumstance. In criticism of Keynes, F. P. Ramsey, five years later, proposed instead a probabilistic measure of consistent belief.[1] But while Keynes at least professed to be continuing the tradition of Bacon and Mill, with its emphasis on experimental variation,[2] Ramsey's theory does not concern itself at all with the typical structure of experimental evidence—the results of manipulating relevant variables. Keynes's theory was designed, like Bacon's and Mill's, to deal with the problem of inductive support for universal hypotheses. But Ramsey's theory could not deal with this problem because it was intended to conform, broadly, to the criterion of adequacy constituted by appeals to betting practice.[3]

It can easily be seen that, if we wish to measure a reasonable man's belief by proposing a bet and then seeing what are the lowest odds he will accept, as Ramsey suggests, we should not expect to measure thus his belief in any testable generalisation over an unbounded domain. For a reasonable man will not accept an unfair wager. But if two people, *a* and *b*, are to make a fair wager with one another, they must at least be able to describe both some circumstances in which *a* will be said to have won the wager and *b* to have lost it, and also some circumstances in which *a* will be said to have lost the wager and *b* to have won it. Both sides must have some conceivable chance of not only winning the wager but also of knowing that they have won it. Now suppose *a* bets on the false-

[1] In the essay 'Truth and Probability', published posthumously in *The Foundation of Mathematics* (1931).

[2] Cf. *A Treatise on Probability* (1921), p. 265 ff. Keynes fails to mention J. F. W. Herschel's contribution to this tradition.

[3] Op. cit., p. 172 ff.

hood of H and *b* on its truth. If H is a singular statement of observable fact, e.g. that the sun will rise at 7.15 a.m. tomorrow, it is easy enough to draw up the terms of the wager in such a way as to allow both *a* and *b* a conceivable chance of knowing who has won the wager. But if H is a testable generalisation over an unbounded domain there is no conceivable chance of anyone's knowing that *b* has won his bet. No-one can both know the truth of each singular conditional deducible from H and also know that these are all the singular conditionals so deducible. People may perhaps believe that H and its consequences are well-supported by the evidence. But that is not enough. For what is at stake is how to measure an evidentially supported belief, and what is offered as a paradigm is the lowest odds a reasonable man would take for a wager that the well-supported belief is true. So it is essential that one should be able to know the truth of the hypothesis independently of the fact that it is evidentially well-supported.

It follows that appeals to betting practice are quite out of place as a criterion of adequacy for theories of inductive support. They are inapplicable to the type of hypothesis normally involved. Admittedly, some philosophers who invoke the betting analogy have supposed that inductive logic is not essentially concerned with generalisations over an unbounded domain. Carnap, for example, has argued that when an engineer, say, holds that a law is very reliable he does not mean 'that he is willing to bet that among the billions of billions, or an infinite number, of instances to which the law applies there is not one counterinstance, but merely that this bridge will not be a counterinstance, or that among all the bridges which he will construct during his lifetime there will be no counterinstance.'[1] But, though this may be perfectly true of anyone like an engineer, navigator, physician or plant-breeder, who wishes to make practical use of a natural law, it is not at all true of those who wish to use one such law for the explanation of others. Scientists need to be able to assess evidential support for universal hypotheses both in establishing the facts to be explained (e.g. elementary, causal or correlational uniformities) and also in establishing explanatory theories. No-one could have supposed that Newton's mechanics explained

[1] *Logical Foundations of Probability*, p. 572. Cf. p. 236 f., and also F. P. Ramsey, op. cit., p. 184. I. Hacking, *Logic of Statistical Inference* (1965), p. 216, mentions, without endorsing, an alternative possible defence of the betting model: 'You might ... imagine yourself pretending to bet on theories and hypotheses as if you were gambling with an omniscient being who always pays up if you are right'. But the object of the betting model is to be able to judge the merits of rival theories in the rather abstruse and perhaps speculative field of inductive logic, by reference to a criterion of adequacy that invokes a familiar, everyday operation; and pretences about wagers with omniscient beings are not an operation of this kind. In any case, so far as gambling involves predicting under conditions of uncertainty, omniscient beings cannot gamble.

the revolutions of the planets if he did not also suppose that all Newton's axioms of motion were more or less universally valid. Moreover, this process of subsuming subordinate regularities under more comprehensive ones is integral, as we have seen (§9), to the attainment of higher levels of inductive support. When two natural uniformities admit of the same theoretical explanation they can thereby corroborate one another. If the betting analogy forces us to dissociate inductive logic from the needs and consequences of theoretical explanation, it has no application to inductive support. There may well be some quite valuable uses for measures of consistent belief, like Ramsey's, or range-measures of logical probability, like Carnap's. But their invocation of the betting analogy serves to highlight the fact that they cannot be used as measures of inductive support, if inductive support is to be construed in terms of the results of experimental tests on universal hypotheses under the manipulation of relevant variables.

Jean Nicod, writing[1] at about the same time as Ramsey, came to see some of the difficulties involved in constructing a probabilistic theory of inductive support by variation of circumstance. In particular he pointed to the fact that assessment of this support always relies on previously acquired beliefs about the operation of certain natural variables. But, instead of arguing therefore that some non-probabilistic theory of experimental support was needed, he concluded that only enumerative induction was capable of giving scientific hypotheses a more than mediocre probability. Instead of first looking at the principles of scientific reasoning in order to see whether or not they admitted of a probabilistic formulation, he adopted *a priori*, as it were, a probabilistic constraint on any admissible measure of scientific reasonableness and then rejected any type of philosophical theory that did not conform to that constraint. Not surprisingly the criterion of confirmation that results from this procedure gives rise to several paradoxes, discussed in §§11–12, when confronted with the realities of scientific reasoning. It is difficult to find a better example of philosophers' first creating a dust and then complaining that they cannot see.

In some ways Karl Popper's *Logik der Forschung* ran counter to the prevailing trend. It emphasised the importance of falsification rather than of confirmation, and of tests rather than mere observations; and it measured corroboration by a logical improbability rather than by a probability. Moreover the preference it gave, other things being equal, to the more general, or comprehensive, of two universal hypotheses seemed to accord with Bacon's conception of a hierarchy of axioms. But

[1] Cf. 'The Logical Problem of Induction', written between 1921 and 1924, and published posthumously in *Geometry and Induction* (1930).

in preferring a probabilistic basis for his measure of corroboration Popper resembled most other writers of the period. Moreover, though a metric for the logical improbability of a hypothesis must to some extent depend on empirical knowledge according to Popper,[1] the corroborative value of an experimental test, as such—i.e. as a test to which more than one hypothesis can be submitted—, is not open to graduated readjustment by unfavourable experience, in Popper's theory, much as the choice of confirmation-function in Carnap's theory is not amenable to empirical control. Yet in actual science the grade of support[2] to be attributed to a hypothesis in the light of a certain test-result must depend on the empirically ascertained relevance of the variables manipulated in the test. Later events must be able to show, as in the thalidomide case, that though the reports of a certain test-result were quite correct we were nevertheless wrong to suppose that they gave so much support to the hypothesis in question. Similarly, even if two correlational generalisations, or two scientific theories, have survived the most thorough tests we know how to devise, the more general, or comprehensive, of the two is by no means always better supported. This is obvious enough if the hypotheses are materially dissimilar to one another, since we inevitably lack any empirical backing for mutually commensurable assessments of support in such a case. But it is also true if the two hypotheses are materially similar to one another. The conjunction and equivalence principles combine to generate a consequence that is demonstrably quite inconsistent with the view that of two successfully tested and materially similar hypotheses the more general one is always the better supported. Suppose two mutually independent universal hypotheses $(x)(\mathrm{A})$ and $(x)(\mathrm{B})$ have both survived their tests, and $(x)(\mathrm{A})$ is at least as well supported as $(x)(\mathrm{B})$. Then, by the conjunction principle—i.e. (8) of §8—$(x)(\mathrm{A})$ & $(x)(\mathrm{B})$ has the same support as $(x)(\mathrm{B})$, and so by the equivalence principle—i.e. (1) of §2—$(x)(\mathrm{A}$ & $\mathrm{B})$ has the same support as $(x)(\mathrm{B})$, though $(x)(\mathrm{A}$ & $\mathrm{B})$ is certainly more general than $(x)(\mathrm{B})$.

In fact, in virtue of its conformity to the general conjunction principle

[1] *The Logic of Scientific Discovery*, p. 404.

[2] Admittedly, Popper did not claim to be offering a measure of evidential support in the original, 1935 edition of his book. But in the appendices to the 1951 translation he claimed quite expressly that what he called degree of corroboration was a measure of 'acceptability' (*The Logic of Scientific Discovery*, pp. 388, 392, 394, 399, 415, 419) or of 'the intuitive idea of degree of support by empirical evidence' (p. 393, cf. pp. 395–6, 399, 410). It is also true that what Popper called the method of 'the sub-class relation' (ibid., p. 115 ff.) is a little similar to some of the non-standard inductive functions discussed in §16 below: cf. p. 154 f. But as such it cannot be a criterion of support. By confounding together questions about acceptability with questions about support Popper contrived greater plausibility for his incorrect thesis that the support for a successfully tested hypothesis varies with its generality—since this generality is obviously important for acceptability (cf. §1, p. 10 f. above, and §16, p. 154 below) though irrelevant to grade of support.

(viz. (13) of §8), grade of inductive support cannot be a function of any relevant probabilities at all, as we have seen in §3. So the present book may be regarded, from a historical point of view, as an attempt to develop Bacon's seminal ideas in a way that is not vitiated by obsession with the mathematical calculus of probabilities. This is not to deny the immensely important role of probabilities within science. Indeed, I have already (§13) described ways of integrating some of the statistical procedures actually used in contemporary science with a gradation of inductive support that derives from the actual procedures of experimental tests. But the present book does not attempt to submerge the differences between inductive support, on the one hand, and statistical estimation, on the other, by creating a wholly new, philosopher's concept of confirmation, like Carnap's,[1] that is intended, ideally, to replace both. Correspondingly, it rejects the view, held by Carnap[2] and many others, that induction by variation of circumstance and so-called enumerative induction should be regarded as coordinate species of a single logical genus.[3] These two modes of reasoning do not share a common logical syntax, nor do they differ only in that the former relies on variety of evidential instances, the latter on number. Nor are they in any systematic sense rivals to one another, as philosophers have often thought they were, since there is no overarching measure of credibility or confirmation which can sit in judgement upon them.

Enumerative induction is, as was argued in §12 above, a special case of one kind of magnitude-estimation, viz. where the magnitude to be estimated is a probability and the probability is equal, or very close, to one. If enumerative induction is to be classified at all, this is the classification that best reveals its underlying logical structure. Inductive support-functions, however, have a quite distinctive logical syntax, and if we wish to treat them as a species of some wider logical genus their genus is best defined in terms of these syntactic principles. I shall henceforth refer to any monadic or dyadic function that satisfies the same logico-syntactic principles that experimental support-functions satisfy, as an inductive function. By examining some non-standard inductive functions, in the following chapter, I shall aim to shed more light on the family of thought-patterns to which reasoning about experimental support belongs, and thereby to fortify my account of its characteristic structure. At the same

[1] Cf. *Logical Foundations of Probability*, Chs. iv and ix, and esp. pp. 518–9.

[2] Ibid., p. 575.

[3] This does not exclude the possibility that they should be regarded as co-ordinate species of a genus for heuristic purposes or for the purpose of achieving psychological explanations of learning. But it looks as though a learning-device that was responsive to repetitiveness of stimulus would have to be differently structured from one that was responsive to variety of experience, even if the two might be capable of co-ordinated functioning.

time I shall aim to illumine the extent of the difference between the family of thought patterns to which this experimental reasoning—i.e. Baconian induction—belongs, and the family to which enumerative induction belongs.

§15. The concept of information

A probabilistic theory of induction has sometimes been used as a basis for explicating the ordinary concept of information. But, though they correctly distinguish the ordinary concept of information from the communication engineer's concept, Carnap and Bar Hillel's explications of the former concept are nevertheless incorrect. By assuming that amount of information is always assessed against a background of existing information they confound 'information' with 'fresh information', and by ignoring the relativity of this amount to a particular problem or question they confound 'information' with 'semantic content'. In fact amount of information may be a function of some statistical measure, or it may be a function of inductive support. But in both cases the appropriate concept of information is syntactically as well as semantically distinct from the homonymous concept in signal transmission theory.

Any clarification of the logical syntax of experimental support, and of its relation to the probability-calculus, is bound to shed light also on an important concept of information which has sometimes been radically misconstrued. This concept is the one familiarly employed when we say such things as 'We need more correct information about the behaviour of the new alloy in stress conditions' or 'Full and correct information is given here from which to estimate the therapeutic efficacy of penicillin' or 'We haven't got enough sound information to go on'. This is, of course, a relatively wide sense of the word, in terms of which we can also define a narrower sense, equivalent to 'correct information' and contrastable with 'misinformation'. But it is in general more profitable to analyse first the wider and weaker sense of such a word, rather than the stronger and narrower, since the latter can subsequently be analysed in terms of the former, plus appropriate qualifications, while the former is not open to analysis in terms of the latter.

Now, there has long been an obvious temptation to confuse the ordinary concept of information with the measure of signal transmission adopted by communication engineers that has very commonly gone under the same name. In the latter sense the information-content of a signal is conceived of as the signal's selective potential. Receipt of such information gives the recipient power to select between the alternatives constituting the domain of his doubt, so that the greater the prior improbability of the ultimate selection, the more information the signal conveys. The fundamental measure for this information is normally the number of binary choices that the signal authorises. If the prior probability of a certain signal's selection is p, its information-content, in the technical, Shannon-Weaver sense, is thus $-\log_2 p$.

But it is easy to see, and has often been pointed out, that the amount of information, in the ordinary sense of the word, that is conveyed by a signal is normally quite independent of its quantity of information in the Shannon-Weaver sense. For example, it may be that a coast-guard's code-book takes whole sentences as its units, and codes each of them as a single symbol for signalling purposes. But the signal for 'A Panamanian tanker of approximately 50,000 tons is three miles offshore' may have occurred twice as often, in the past, as the signal for 'A ship of unidentified type and nationality is offshore'. So the latter signal has twice the improbability of the former and therefore much greater information-content in the Shannon-Weaver sense. But in the ordinary sense, to anyone interested in the shipping situation, it may well convey much less information. Hence most of the time, though with occasional lapses, writers on the theory of signal transmission[1] have been careful to insist on the difference between their measure of a signal's so-called information-content, which applies to the coded form of the signal irrespective of any linguistically or conventionally assigned meaning that the signal may have, and the ordinary concept of the information conveyed by a signal, which is unknowable without a knowledge of the signal's meaning.

Indeed, it has sometimes been held that the quantity of information, in the ordinary sense, that is conveyed by a statement has intrinsically nothing to do with communication,[2] and it has been argued instead that this quantity is a function of the statement's logical probability. Thus Carnap and Bar-Hillel[3] have offered two possible explications for the ordinary concept of information. One, called 'inf', is such that where the functions denoted by 'inf' and 'c' are based on the same Carnapian range-measure, $\mathrm{inf}(H/E) = -\log_2 c(H, E)$. That is, the amount of information contained in a new message H with respect to existing knowledge E tends to increase logarithmically with the relative inductive improbability of a hypothesis H on the evidence E. The other explicans, called 'cont', is such that, where the functions denoted by 'cont' and 'c' are based on the

[1] E.g. C. Cherry, *On Human Communication* (1957), p. 50. For an example of a lapse, cf. N. Wiener, *The Human Use of Human Beings* (1950), p. 7 f.

[2] E.g. Y. Bar-Hillel, *Language and Information* (1964), p. 287 (in 'An Examination of Information Theory', reprinted from *Philosophy of Science* xxii, 1955, p. 86 ff.).

[3] Their ingenious and important paper 'An Outline of a Theory of Semantic Information', was first published as Technical Report no. 247 of M.I.T. Research Laboratory of Electronics (1952), and subsequently reprinted in Y. Bar-Hillel, op. cit., p. 221 ff.

same Carnapian range-measure m, cont(H/E) = m(E) × (1 − c(H, E)). So that again the amount of information contained in H with respect to existing knowledge contained in E increases with the inductive improbability of H on the evidence E. Carnap and Bar-Hillel compared these two measures in regard to additivity and suchlike mathematical or logical properties, and claimed that each measure is intuitively plausible in some respects and implausible in others. They therefore concluded that there is not just one ordinary concept of information but at least two. They also noted that, where E is tautological and accordingly inf(H/E) = inf(H), we get the result that $\inf(H) = -\log_2 m(H)$. So, since the range-measure in m(H) is the (absolute) logical probability of H, the syntax of the first of the two Carnap-based functions of semantic information may be mapped on to the Shannon-Weaver theory of signal transmission. On this view the ordinary concept of information has the same logical syntax, though not the same semantics, as the technical concept of information employed by communication engineers.

Unfortunately, whether or not these Carnap-based functions explicate any familiar concept at all, they come nowhere near explicating the ordinary concept of information. Though there is certainly an important relation between that concept and the concept of inductive support, it is not the kind of relation that Carnap and Bar-Hillel suggested. But an examination of where their theory goes wrong will help to show what the situation is really like. Specifically, the theory goes wrong in two main ways—first, by ascribing to the ordinary concept of information a feature which it has not got, and secondly, by failing to ascribe to it a feature which it has got.

First, then, the theory implies that in assessing the amount of information afforded by a statement we must always measure this relatively to some other statement. Both 'cont' and 'inf' are dyadic functors, and the interpretations offered for them imply that quantity of information is always assessed against a background of existing knowledge. Not that a Carnap-based theory is incapable of representing an absolute assessment. But it achieves such a representation only by taking a degenerate case of background knowledge, viz. where E in 'inf(H/E)' or 'cont(H/E)' is a tautology. In normal contexts, however, the term 'information' does not function like this. We normally hope that our informants will provide us with fresh information, but when they tell us what we know already they provide us with stale information, not with no information at all. We say 'That's no longer news to us', not 'That's no longer information to us'. The statement 'Information was given to the students by the physician about the therapeutic efficacy of penicillin' is not rendered false by the fact that the students already knew of this efficacy. Similarly, if they already knew something about it and wanted to know more, they would ask for *more* information. If they asked instead just for information about it, the physician would assume that they had not heard, or had not understood, what he was saying.[1]

Someone may well object here that, even if they were offered as explications of the ordinary concept of information, *tout court*, the functions proposed by Carnap and Bar-Hillel could perfectly well be taken instead as alternative explications of the ordinary concept of fresh information. But these functions have a second inadequacy, which bars any such escape-route.

Though the ordinary concept of the amount of information imparted by a statement is not, *pace* Carnap and Bar-Hillel, relative to another statement about what is already known, it is certainly relative to a question or problem of some kind, and to that extent it has something to do with communication. When we say that the bystanders gave the policeman all the information they could, we imply that they told the policeman everything they knew about the accident or crime he was investigating, not that they told him everything they knew about the weather, their hobbies, their children's health, latest football scores, and so on. The query 'Have you any information?' calls for the counter-query 'About what?', since every normal person has information about something. No-one can compare the amount of information imparted by one statement with the amount imparted by another unless he knows what kind of problem requires solution. If we are interested in to-morrow's weather there is more information to be gained from one barometer reading today than from the whole of Gibbon's *Decline and Fall of the Roman Empire*.

Not that the only proper account of the ordinary concept of information has to be a pragmatic or psychological one, whereby assessments of amount of information are relativised to a person at a time. This is no more needed in the case of information than in that of explanation or support. In order to study the logical structure of such concepts it is convenient to abstract from the pragmatic or psychological dimension of analysis and confine the domain of the appropriate functors, or relational predicates, to propositions, rather than require them to have one argument-place that is filled by expressions denoting people-at-a-time. We can talk profitably about the conditions under which a scientific

[1] The adjective 'informative', however, is a little more specific on occasion than the noun 'information', since we might be inclined to call 'uninformative' a report that was full of stale information. Adjectives tend sometimes to be more specific, or evaluative, than the corresponding nouns (cf. 'cost' and 'costly', or 'shape' and 'shapely'), perhaps because it is often more difficult or more awkward to qualify an adjective with an adverb than a noun with an adjective.

theory explains a correlational generalisation, or an evidential report supports a hypothesis, without having to suppose that we are talking about the effect of learning something on some person's mind. Similarly we can talk about the extent to which a certain proposition affords information that solves a certain problem, without having to suppose that we are talking about the extent to which a query in someone's mind has been answered.

When philosophers neglect this relativity to problem or interest in the ordinary concept of information, as do Carnap, Bar-Hillel and many others,[1] the explicandum they have in mind seems closer to 'semantic content' than to information. It is as if they had correctly seen that why the communication engineer's theory of signal transmission was inapplicable to the ordinary concept of information was because it took no account of a signal's meaning, and then incorrectly concluded from this that an adequate theory of information must explicate the concept of meaning or semantic content. It is as if a feature that is necessary, in order to constitute the explicandum, had been taken as sufficient to constitute it.

But information is very different from semantic content. People often seek information about matters that concern them but it does not make sense to speak of their seeking meaning or semantic content. Indeed giving information to others is one of the commonest purposes with which people open their mouths. But it is far from being their purpose on every speech-occasion. Telling others what to do, amusing them, making enquiries, etc. are other, co-ordinate kinds of purpose. Meaning, on the other hand, is something that all speech has. Also, information can be false as well as true, like the statements that impart it. But meaning, or semantic content, is neither. Information therefore can guide or mislead, be valuable or valueless, but meaning, as such, cannot.

The only way, it seems, in which the theory proposed by Carnap and Bar-Hillel could begin to accommodate the relativity of information to problem or interest, is on the assumption that 'inf' and 'cont' may be defined separately for an indefinitely large number of formalised languages, each of which differs from the others in some aspects of its vocabulary (in particular, the semantical rules for individual and predicate constants) and represents through these differences a different focus of intellectual interest. But even then there would be no statement,

[1] E.g. D. M. Mackay, 'Quantal Aspects of Scientific Information', *Philosophical Magazine* xli (1950), p. 289 ff., and J. Hintikka and J. Pietarinen, 'Semantic Information and Inductive Logic', in *Aspects of Inductive Logic*, ed. J. Hintikka and P. Suppes (1966), p. 96 ff. For a connection between inductive logic and the assessment of meaning, as distinct from information, see §16, p. 143 ff.

on each particular occasion, of the actual problem in relation to which we were assessing the amount of given information. That problem would be represented in only a very vague and inexact way by the semantical selectivity of the language. Moreover, the more selective the vocabulary of each language was made, in order to determine the designated foci of intellectual interest more narrowly, the less opportunity there would be to make comparisons between the information given by different statements in relation to the same problem or by the same statement for different problems.

All these difficulties can easily be overcome, for an important class of cases, if one takes the information-carrying proposition in the inductive situation to be not the hypothesis but rather the evidential proposition. A dyadic support-functor, denoting the extent to which E supports H, is also some gauge of the information that utterance of E would impart in relation to the question whether H is true. Admittedly, we can deal in this way only with information in the sense of 'evidential information' —i.e. information from which one can infer or estimate the correct answer to a question, rather than information that actually constitutes such an answer. But it is only where the information is thus partial or incomplete that assessment of how much information one actually has presents any problems. Moreover, it turns out to be quite plain that a function which assesses information in this sense cannot be mapped on to the homonymous feature of signal transmission theory. It is syntactically as well as semantically distinct from the latter.

Three different types of case need to be mentioned here. First, we can talk of a large evidential sample's providing more information than a small sample—as an answer to the question: what is the real magnitude? —in that it enables us to make the same estimate of some statistical magnitude with, say, a higher coefficient of confidence or a closer approximation (more accurate estimate) with the same coefficient of confidence. Of course, some small samples from a particular population will match the population as a whole much better than some larger ones. There is thus a sense in which these better-matching samples could be said to give us more valuable information. But a larger sample will normally give us *more* information, in the sense that, at a given confidence-level, it imposes a greater restriction on the interval within which the magnitude in question must be located. What is also to be noted is that though this information-measure is statistical in character its syntax cannot be mapped on to signal transmission theory. As sample-size increases, the increase in information thereby provided is measured by the increase, not the decrease, in the coefficient of confidence (cf. §12,

p. 109 ff.) for the same estimate. Quantity of information thus varies directly with the appropriate probability, not inversely, as in the case of the transmission-theoretic concept of information.

Secondly, there is the standard type of evidential information that is gained from experimental tests. But the appropriate inductive support-function cannot itself be regarded as an information-function. For, if the question at issue is whether H is true or false, we get just as much information from E if E supports not-H to such-or-such an extent, as if E supports H to that extent. We shall therefore want to define any inductive information-functor i[E, H], denoting the grade of inductive information afforded by E in relation to the question whether H is true, in such a way that where we have s[H, E] > s[not-H, E] we get i[E, H]= s[H, E] and where we have s[H, E] < s[not-H, E] we get i[E, H]= s[not-H, E].[1] A suitable set of definitions will be proposed in §22 below, and some consequential theorems will be developed. All that needs to be pointed out here is that since inductive support is not a function of any relevant logical or statistical probability, the logical syntax of inductive information cannot be mapped on to any theory, like signal transmission theory, that treats information as a function of one or more probabilities.

Thirdly, there may be other kinds of general questions than those that evidential samples or experimental tests may help us to answer, e.g. questions about what is morally or legally proper to do. Support for answers to such questions may sometimes be inductive in its logical syntax, as we shall see in §17–18 below, and the premisses that afford this support may accordingly be said to supply inductive information. So an inductive information-functor can take certain propositions as fillers of its first argument-place that are not reports of experimental test-results.

In this context it is also worth mentioning that to preserve syntactic continuity with signal transmission theory, writers have sometimes suggested, or hinted, that a measure of evidential information may be gained by a reinterpretation of the Bayesian measure of communication in the presence of noise.[2] Where X is the signal as actually transmitted and Y is the noisy received signal the transmission-theoretic information-

[1] This kind of definition could be generalised to cover questions that do not admit of the answers 'Yes' or 'No' (as does the question whether H is true) but involve the use of interrogative prefixes, like 'which', 'when', 'how many', 'how long', etc. If H describes such a question, and if our logic can determine the range of admissible answers to this question, then i[E, H] would need to be equated with the highest grade of support that E gives to any member of that range. But no further issues seem to be raised thus that are of primary interest for the logic of induction as distinct from the logic of interrogation.

[2] E.g. C. Cherry, op. cit., p. 200 ff.

content of Y is given as

$$(3) \quad \log_2 \frac{p[X, Y]}{p[X]}$$

So what (3) measures, in this transmission-theoretic interpretation, is the amount of successful transmission that has taken place irrespective of what the meaning of the message was. But if instead we treated X and Y in (3) as propositions we might appear to have a measure of the amount of evidential information that Y gives us in relation to the question whether X is true. With apparent plausibility this amount would grow larger as the relative, or *a posteriori*, probability of X on Y increased, and smaller as the absolute, or *a priori*, probability of X increased. It might thus appear to be as if nature could be treated as the transmitter and experimental science as the receiver.

However, these appearances are deceptive. For we come up here against the familiar logical difficulty, as in Carnap's or Popper's theory, of scientific reasoning, of how to determine the absolute, or *a priori* logical probability of a proposition. There is a clear, agreed sense in which the absolute, or *a priori*, probability of X in (3) can be determined if 'X' is interpreted as denoting a particular type of transmitted signal. This absolute probability is X's frequency of occurrence in the population of transmitted signals. But there is no suitable clear and agreed sense in which the absolute, or *a priori*, probability of a scientific hypothesis can be determined. To base the measure of such a probability upon the choice of a language, language-system, or semantical model—which is the usual proposal—is altogether too arbitrary, as we have already seen in relation to the problem of measuring relevance (§6).

V

Some Non-Standard Uses of Induction

§16. The grading of evidentially permissible simplification

In the face of unfavourable test-results we may prefer to adjust the meaning, or formulation, or domain of application, of the hypothesis, rather than assign it a low grade of evidential support. Though such adjustments may be of one or other of these three types, a single kind of function will serve for the assessment of all evidentially enforced complications or evidentially permissible simplifications. For we can define a normal hypothesis' grade of inductive simplicity as dependent on the number of relevant variables of which it omits to make explicit or implicit mention (or the number of relevant variables of which it omits to mention variants), and correspondingly a non-tautological hypothesis of maximum complexity becomes capable of achieving full support from the outcome of a single trial. In this connection it is important to note that what appears as an idealisation, in comparison with the general run of nature, is bound to be a complication from the point of view of inductive logic, since the operation of one or more relevant variables is excluded. Assessments of evidentially permissible simplification are true if and only if corresponding assessments of evidential support are true, but do not say the same as the latter.

Theories of inductive support have been attacked by some philosophers on certain historical, or quasi-historical, grounds. Such a philosopher stresses the point that experimenters do not by any means always reject a hypothesis when it has been falsified. If an observed phenomenon falsifies an otherwise well supported hypothesis, experimenters may prefer to treat this phenomenon just as an exception which lies outside the scope of the hypothesis in question. Therefore, the argument runs, one should not suppose that any theory of inductive support, assessing the merit of a hypothesis in terms of its successful or unsuccessful trial-outcomes, can tell the whole story about experimental reasoning in natural science.

This is an important point. But it should be treated as implying a criterion of adequacy for any theory of inductive support, not as indicating a deficiency from which such a theory must inevitably suffer. An adequate theory of inductive support will be so constructed that it is quite consistent with the normal kinds of non-standard responses to falsificatory evidence; and several of these exist.

Two kinds of situations do not, however, concern us here.

First, a single trial-outcome may have been unfavourable on a particular occasion, but on repetition of the test perhaps the same trial's outcome is no longer unfavourable. We have already dealt with the principle governing such a case, in §8. The difference in outcome between the two occasions must be put down to the hidden operation of some relevant variable that is not openly and expressly manipulated in the test. So the evidential propositions that report these outcomes cannot be true reports of canonical test-results.

Secondly, if a certain number of unfavourable trial-outcomes keep on recurring, though irregularly, it may be that instead of trying to find support for a hypothesis asserting that anything, if an R, is an S, a researcher would do better to try and find support for a hypothesis asserting that the probability of an R's being an S is within a certain interval of some suitable figure. We have however already (§13) shown how our theory of inductive support can apply to such statistical hypotheses, and nothing further needs to be said about them here.

But an unfavourable trial-outcome that is believed, or assumed, to be regularly repeatable calls for rather different treatment. The standard response is to say that the hypothesis cannot pass a test of more than a stated degree of thoroughness, and that it does not therefore have more than the corresponding grade of support. But this response assumes that the meaning of the hypothesis is fixed and the grade of support remains to be discovered. We might instead prefer to assume that full support has in any case to be established and that the meaning which the hypothesis can have, if it is to achieve this grade of support, remains to be discovered. Clearly, if we chose to speak like this, we should need to individuate hypotheses somewhat differently. Different sentences might express the same hypothesis, in this sense, even if they varied slightly in meaning. But the various gradations of meaning that were possible would be closely related to the conditions under which the various grades of support were achieved or missed. We may accordingly hope to construct inductive functions that will, in effect, assess these gradations. Let us call such a function a 'simplification-function', since it assesses the extent to which the evidence allows an experimenter to simplify his hypotheses, or forces him to complicate them.

The meaning of a hypothesis may be adjusted in different ways. A term or terms in it may be redefined; a condition may be added to, or taken away from, its antecedent; a restriction may be imposed on, or removed from, the characterisation of its domain of discourse; or some combination of these methods may be employed. Nor is the choice of method arbitrary. In practice there may be good reasons why, on a particular occasion, an experimenter prefers one method of simplification, or complication, to another. For example, in a relatively elementary, first-order generalisation it may be best to conjoin an additional condition with the antecedent, since this is the most obvious way in which the need to modify meaning can be acknowledged. But a correlational generalisation is perhaps best modified by an adjustment to its domain of application: a neat mathematical formula is often more easily incorporated into a comprehensive theory than a more complicated one is, and if its domain has to be restricted in some way in order to permit such neatness it is easy enough to transfer this restriction to the theory.

Indeed, these are the terms in which we can best understand the process of idealisation, as it is called, that often needs to take place when a single correlational generalisation is constructed to embrace a multitude of more elementary hypotheses. Typically, a domain of individuals is postulated that are not subject to the operation of whatever variables would otherwise create exceptions to the generalisation. For example, if the generalisation correlates the velocity of a falling body with the period for which is has been falling, it may be necessary to confine the domain of the generalisation to objects in a frictionless medium. Perhaps the velocities obtained in actual experiments, where friction operates, cannot be obtained as any straightforward function of period of fall. So they must be explained as friction-caused deviations from the velocity under ideal conditions. But what from the point of view of the real world is an idealisation, because there is in fact no frictionless medium, must be seen, from the inductive point of view, as a restriction of meaning, because it is the price that has to be paid for freedom from a particular kind of unfavourable test-result. The domain of the generalisation is a simpler one than really exists, but its characterisation in familiar terms is correspondingly more complex since certain features of objects as we know them—certain effects they are liable to undergo—have to be excluded. The generalisation tells us what would happen under ideal conditions and so appears to correct the facts, as it were, by describing how nature should behave, rather than how it actually behaves. But, in order to achieve this, the domain of the generalisation cannot be characterised without some allusion to every variable that is supposed to be responsible for the counter-evidence actually found in the ordinary run of nature—or without at least some summary description, such as 'under laboratory conditions', of an approximately similar domain.

A somewhat different situation arises in the case of causal hypotheses.

Here the object must be to achieve a hypothesis that has not an ideal domain, unrealised in the actual world, but rather a domain including just those circumstances in which people normally operate the causal factor specified in the hypothesis. For example, the hypothesis 'If a cue strikes a billiard-ball in the middle of a billiard-table, it will cause it to move' is obviously open to falsification by experiments using glue or nailed barriers. But it would be uneconomical to cumber the hypothesis with added conditions about the absence of impediment from glue, nailed barriers, etc. Very many such conditions might need to be mentioned, and in any case none of these impediments are normally encountered in billiard-tables. The most helpful way to specify a causal process is in terms of how it operates on normally encountered circumstances. So the best place to adjust the meaning of the hypothesis is again—though for a different reason—in the characterisation of its domain. The hypothesis applies, we say, to 'normal circumstances'.

But although the meanings of different kinds of hypotheses may thus be adjusted, for different reasons, in different ways, there seems no reason why the extent of the adjustment should be assessed by correspondingly different criteria. From an inductive point of view the price thus paid for avoiding a particular kind of unfavourable test-result is always of the same kind. The absence of impediment from one or more combinations of variants of relevant variables has somehow to be written into the hypothesis, whether it be by redefinition of a term, by addition of a condition to the antecedent, or by restriction of the domain of application.

Sometimes, perhaps, where the adjustment is to the domain of application, it does not emerge in ordinary speech as a change of meaning. This is particularly the case with singular substitution-instances of universal hypotheses. It is evident enough that there has been a change when 'If *a* is R, *a* is S' has to be replaced by 'If *a* is R and V, *a* is S'. But how is the change to be represented for such a singular statement when the domain of the generalisation, not its antecedent condition, has been modified? Certainly anyone who says 'If you strike this billiard-ball with your cue, you will cause it to move' implies that all the relevant circumstances are favourable. But he does not imply which circumstances these are. It is not normally part of the meaning of his statement, that there is no impediment from glue or nailed barrier on the billiard-table. Correspondingly even if he learns more about the factors that may obstruct the ball's movement, he can repeat his statement with unchanged meaning. What he has learned can be presumed to belong to the conditions for the truth of the statement, rather than to its proper meaning. Nevertheless conditions of truth do often get absorbed into meaning,[1] and in order to avoid unrewarding complexities in our philosophical reconstruction of the situation it is convenient to assume that this is what happens here.

If, therefore, we accept an appropriate equivalence principle, it looks as though a single type of simplification-function will serve normal scientific purposes. But what should we take to be the characteristic criteria for such a function's value-assignments, and what therefore is its logical syntax?

I assumed originally (in §5, p. 41) that a hypothesis to be tested will not contain a term describing any relevant variable or variant of a relevant variable. This restriction was relaxed somewhat, when I came to take correlational generalisations into account, so that test t_1 on a testable hypothesis U could be regarded as manipulating the variable described by the antecedent of U (or some finite sub-set of it). That variable was then to be taken as the first in the ordered set of variables $v_1, v_2, \ldots v_n$ that are inductively relevant to U. But even with this relaxation it would follow that if a hypothesis did contain a term that described a variant of a relevant variable v_i, where $i > 1$, the hypothesis might on that account alone have zero-grade support according to the support-function appropriate to these hypotheses, whatever the evidence. And this somewhat unrealistic assumption, made for clarity of exposition, is hardly compatible with taking a serious interest in the different grades of simplification that may be allowed by given evidence or the different grades of complication that may be enforced by it. I shall therefore now completely abandon the assumption that a hypothesis to be tested will not contain a term describing any relevant variable or any variant of a relevant variable. I shall continue to define a particular class of materially similar hypotheses by reference to a particular category of non-logical terms, and to define a particular support-function by reference to such a class. The variables relevant to such a class of hypotheses, and their order of relevance, are to be determined as before, and no expressions describing variants of these relevant variables are to be, or to be definable in terms of, members of the category of non-logical expressions defining that particular class of materially similar hypotheses. But I shall now call hypotheses in which all the non-logical expressions belong to this category unmodified hypotheses; and I shall extend the concept of a class of materially similar hypotheses that can have positive support, according to a

[1] Cf. L. Jonathan Cohen, *The Diversity of Meaning*, 2nd ed. (1966), p. 156 f. On the systematic equivalence of any set of generalisations over n different domains to a corresponding set of generalisations over a single domain, cf. Hao Wang, 'The Logic of Many-Sorted Theories', *Jour. Symb. Log.* xvii (1952), p. 105 ff.

particular support-function, so as to include within the class certain hypotheses that mention relevant variables or variants of them. These hypotheses will be spoken of as 'inductively modified versions' (or 'versions', for short) of certain other propositions, and each such version will be spoken of as having a certain ordinally numbered 'grade of inductive simplicity' (or as being an '*i*th grade version', for short). Hypotheses that are capable of acquiring inductive support from the results of experimental tests in the normal way will turn out to have a full series of versions, viz. where there are n possible grades of support, a 0th grade version, a set of 1st grade versions, a set of 2nd grade versions, ... and an nth grade version. But logically true propositions will turn out to have only nth grade versions, and all other propositions only 0th grade ones.

Specifically, where there are n possible grades of positive support (because there are n relevant variables cf. §9, p. 80), any testable and unmodified proposition is to be termed an nth grade version of itself and its equivalents. If $n > i > 0$, an ith grade testable version of a testable and unmodified proposition U, or of any of U's equivalents, is obtained from U by conjoining with the antecedent of U some allusion, in order of supposed relevance, to each of the n–i least relevant variables. This allusion may be made in one or other of three different forms. In the first form it is made by assuming that a certain specified combination of variants of the variables is present, as where

(1) Anything, if it is R and V_{i+1} and V_{i+2} and ... and V_n, is S

is an ith grade version of

(2) Anything, if it is R, is S.

For example:

Anything, if it is a hare and not in snow, is grey

would be an n–1th grade version of

(3) Anything, if it is a hare, is grey,

if weather (i.e. snow/not-snow) is supposed to be the least relevant variable for hypotheses materially similar to (3). In the second form the allusion may be made by assuming that any variant or combination of variants of the n–i least relevant variables is causally inoperative, as where

(4) Anything, if it is R and not affected by any variant or combination of variants of v_{i+1}, v_{i+2}, ..., v_n, is S

is an *i*th grade version of (2). For example:

> Anything, if it is a hare and not affected by the weather, is grey

might be an *n*–1th grade version of (3). And in the third form an allusion of the type to be found in (4) takes over at the point at which an allusion of the type to be found in (1) leaves off, as in

(5) Anything, if it is R and V_{i+1} and V_{i+2} and ... V_j and not affected by any variant or combination of variants of $v_{j+1}, v_{j+2}, \ldots, v_n$, is S.

But a 0th grade version of a testable proposition U, or of any of U's equivalents, is always obtained from U by conjoining the antecedent of U with its own denial. In this way

(6) Anything, if it is R and not-R, is S

is a 0th grade version of (2).

We have to think of the modifications of (2) that are evident in (1), (4), (5) and (6) as designed to ensure that every test t_k, if $k > i$, should have a successful result, even in circumstances where normally some variant of a relevant variable, or some combination of variants of relevant variables, would obstruct this happening. Whatever the evidence, every testable and unmodified proposition has some suitably modified version (or versions) that is (or are) proof against falsification. If the evidence gives some positive, but not full, support, the antecedents of (1), (4) and (5) may be thought of as being designed to exclude the operation of the obstructive variant or combination of variants. But no relevant variable, other than that alluded to in the antecedent, is manipulated in a canonical test t_1 (cf. §7, p. 54, and §9, p. 80). Hence no variant of the other relevant variables, or combination of variants of the other relevant variable, may obstruct a successful result there: all such variants or combinations of variants are supposedly inoperative. In t_1 as applied to (2), it is the presence of R itself that may prevent a thing's being S. Hence a version of (2) that is designed to prevent falsification by t_1 must insert into the antecedent an assumption that is designed to exclude the operation of R, with the result that the antecedent becomes self-contradictory and the version so obtained is logically true, as in (6). In other words, if a testable and unmodified hypothesis fails even the least thorough test, t_1, only a tautological version of it can escape falsification. Just as such a hypothesis has 0th grade support, so the version of it that escapes falsification has 0th grade inductive simplicity.

What about other kinds of proposition? If H is a testable and unmodified proposition, a substitution-instance of an *i*th grade version of H must obviously be taken as an *i*th grade version of the corresponding substitution-instance of H or of any of its equivalents; and, in general, any proposition, that is deducible by a certain sequence of (uniquely determining) logical or mathematical operations, or definitional substitutions, from an *i*th grade version of H, must be taken as an *i*th grade version of any equivalent of any unmodified proposition that is got by that sequence of operations from H. If H is a logically true proposition or one that is guaranteed by accepted mathematical postulates, or other non-contingent assumptions, then H has just an *n*th grade version, viz. H itself; for, since such a proposition cannot be falsified, it may conveniently be supposed to have maximum support on any evidence and thus no version of less than maximum simplicity will ever be needed. Finally if H is, or is equivalent to, a contingent conjunction of propositions H_1, H_2, . . ., and H_m, then a conjunction of a version of H_1 with a version of H_2 . . . and a version of H_m, where the lowest grade version is of *i*th grade simplicity, is to be taken as an *i*th grade version of H: we have to think of complexity as a taint with which a complex hypothesis infects, as it were, any conjunction of which it constitutes a conjunct. But if H is not a testable and unmodified proposition, nor an unmodified proposition that is deducible from a testable one nor a logical truth or one that is guaranteed by accepted mathematical postulates or other non-contingent assumptions, nor equivalent to a conjunction of such propositions, then H has only a 0th grade version, which is formed by disjoining H with its own negation. For, since such a proposition is either incapable of acquiring any positive inductive support or is itself an inductively modified version, it needs no inductively modified version of greater than 0th grade simplicity.

For logical purposes no other types of inductively modified versions need be supposed to exist, in the case of first-order generalisations; and for correlational generalisations—like (7) of §9—very similar types of version exist, with corresponding insertions of 'and V_{i+1} . . .' etc. in their antecedents. In practice, as pointed out earlier, any version of H that differs from H in one of the ways described will have an equivalent that differs from H in its domain of discourse instead, or in the definition of one or more of its terms. But the most perspicuous approach to the logic of inductive simplification seems to be in terms of the kinds of version described. Everything is then fully explicit and we need not become involved in side-issues about many-sorted quantification or the structure of stipulative definitions.

Any appropriately chosen *i*th grade version, U′, of a testable and unmodified proposition U, where $i > 0$, can readily be seen to earn positive support in much the same way as U itself earns such support. Suppose, first, that U′ is like (4) and that E is a report of test t_i that gives U ith grade support, where $n > i$. Then the characteristic inclusion (cf. §7) of '$T_i x$' within each of E's existentially quantified reports of U's trial-outcomes ensures that E may be construed not only as the report of a successful test t_i on U but also as the report of a successful test t_n on U′, since E reports, in effect, the co-instantiation of the antecedent and consequent of U′ in each possible combination of variants of relevant variables that can have any effect on U′. Secondly, if E reports a result of test t_n that gives U just ith grade support (cf. §8, p. 60 f.), E may also be construed as the report of a successful test t_n on an appropriately selected ith grade version of U that is like (1), since here too E reports the co-instantiation of the antecedent and consequent of the version in each possible combination of variants of relevant variables. Thirdly, if E reports results of test t_j that suffice to give U just ith grade support (cf. §8, p. 60 f.), where $n > j > i$, then E may be construed as a test-report giving nth grade support to an appropriately selected ith grade version of U that is like (5). But, if U′ is an ith grade version of U and E reports U just to have passed a test t_j where $i > j$, E cannot be construed as giving U′ more than jth grade support, since there would then be at least one more thorough test from which U′ would have something to fear.

Thus there is always some appropriate ith grade version, U′, of a testable and unmodified proposition U such that, for any E, s[U, E] $= i/n$ if and only if s[U′, E] $= n/n$, and the substitution-instances and equivalents of U′ derive inductive support correspondingly, in accordance with the uniformity and equivalence principles. Indeed there is obviously an inverse relationship between the inductive simplicity of a universal hypothesis and the structural simplicity of the test that is required to give it full support. Or, to put the point more exactly, we can say that the simpler U′ becomes the larger is the number of trial-outcomes required for the evidence to give U′ full support, while the more complex U′ becomes the smaller this number can be. For example, if there are just four relevant variables and each has three variants and no variant of any one variable excludes the presence of any other, the number of trial-outcomes required to give full support to a version of 4th grade simplicity is 81. But full support for a 2nd grade version would require only 9 trial-outcomes.

It follows that, if H is an unmodified hypothesis, then for every support assessment stating that, on the evidence of E, H has at least ith grade support, or not more than this, as the case may be, there is an equivalent assessment of evidentially permissible simplification stating that, on the evidence of E, the simplest fully-supported version (or versions) of H has (or have) at least ith grade simplicity, or not more than this, respectively. I.e., where s′ is the simplification-function that corresponds to s, $s[H, E] \geqslant i/n$ if and only if $s'[H, E] \geqslant i/n$, and $s[H, E] \leqslant i/n$ if and only if $s'[H, E] \leqslant i/n$. Similarly, if both H and I are unmodified, then $s[H, E] \geqslant s[I, E]$ if and only if $s'[H, E] \geqslant s'[I, E]$: that is, on the evidence of E, H has at least as much support as I, if and only if on that evidence the simplest fully supported version of H has at least as high a grade of simplicity as the simplest fully supported version of I. Correspondingly a monadic simplification-function assesses the grade of simplicity which the simplest fully-supported version (or versions) of H has (or have); and analogous equivalences hold between $s[H] \geqslant i/n$ and $s'[H] \geqslant i/n$, etc. In each case the simplification-assessment, whether monadic or dyadic, is true if and only if the corresponding support-assessment is true, and this equivalence follows logically from our conceptions of inductive simplicity and inductive support. But the two assessments do not say the same thing,[1] or answer quite the same interest. $s[H] = i/n$ or $s[H, E] = i/n$ tells us what we want to know if we assume that the meaning of our proposed hypothesis is fixed and its grade of support remains to be discovered. $s'[H] = i/n$ or $s'[H, E] = i/n$ tells us what we want to know if we prefer to assume that the hypothesis must in any case achieve full support and that the width of meaning it can have, under this condition, remains to be discovered.

Simplification-functions are thus an important species of non-standard inductive function. They share all the syntactic features of monadic and dyadic support-functions—conforming to the consequence principle, general conjunction principle, and so on—because of the equivalence just mentioned between assessments of evidentially permissible simplification, on the one hand, and assessments of evidential support from test-passing, on the other. They belong to the family of thought-patterns to which experimental reasoning, i.e. Baconian induction, belongs, as distinct from the family to which so-called enumerative induction belongs. Admittedly, one could compare *a priori* requirements about the grade of simplicity that a hypothesis must have (in addition to the requisite grade of support) if it is to be acceptable, with *a priori* requirements about precision, significance etc. in the case of statistical estimation. But whereas confidence intervals, for example, are measured by a probability, grade of evidentially permissible simplification is essentially non-probabilistic because it is an inductive function.

[1] For a discussion of the difference between what a proposition says and what it means, and of the corresponding differences between criteria for sameness of saying and criteria for sameness of meaning, see L. Jonathan Cohen, op. cit., p. 161 ff.

It may be thought paradoxical to suggest that a non-tautologous universal hypothesis should be capable of deriving full support from a single trial-outcome. For this is certainly what is being claimed here in relation to 1st grade versions of elementary hypotheses. A 1st grade version of (2) that is like (4) turns out to be fully supported by an evidential proposition that just reports a successful result of applying test t_1 to (2), viz.

(7) $(\exists x)(Rx \& Sx \& T_1 x)$,

since '$T_1 x$' implies that x is unaffected by any variant of $v_1, v_2, \ldots, v_n$. Similarly, if E just states

(8) $(\exists x)(Rx \& Sx \& V_1 x \& V_2 \& \ldots \& V_n x)$

we need no further statement of evidence than E in order to be entitled to claim full support for a 1st grade version of (2) that is like (1).

But such a situation is not a mere figment of logical imagination. If an experimenter wishes to approximate as closely as he can, under laboratory conditions, to the idealised situations that form the domain of some proposed correlational generalisation, he may well aim at test-reports like (7). In a well-understood field of enquiry, where enough is known about relevant variables to screen off their operation effectively, a single experiment—i.e. a single trial—may suffice to establish an elementary hypothesis. Drop your object in as near a vacuum as you can create, and you can measure its speed of fall, after such-or-such an interval, under as nearly ideal, frictionless conditions as you can get.[1] In the normal circumstances of the billiard-room, a single demonstration suffices to show that a billiard-ball moves when struck by a cue. Even if you cannot screen off the operation of relevant variables effectively you can at least fix on a certain normally encountered variant of each relevant variable, if you know what all the relevant variables are, and support a hypothesis like (1) by an evidential report like (8).

W. E. Johnson coined the term 'intuitive induction' for the process of mind by which a single instance is seen to give full support to a universal hypothesis.[2] In this species of generalisation, he said, we intuit the truth

[1] Of course, to establish an idealised correlational generalisation, as distinct from an elementary hypothesis, by a test t_1 of this nature, we should need to execute more than one trial. Having excluded such factors as friction, magnetism, etc. from the domain of a generalisation (as in §9) about the velocity of falling bodies, we have still to test it under the manipulation of its antecedent variable, viz. after appropriately differentiated periods of fall. Where the domain-conditions for an idealised generalisation are not experimentally approximatable, support for the generalisation can be obtained only by treating it as one of the axioms for a scientific theory, as in §9, p. 84 ff.: among the other axioms will be those generalisations that determine what deviations occur when ideal conditions do not prevail.

[2] W. E. Johnson, *Logic*, pt. II (1922), p. 29.

of a universal proposition in the very act of intuiting the truth of a single instance. He therefore classed it as a form of what he called demonstrative inference, whereby a conclusion can be as certain as its premisses (as in formal-logical deduction), and he contrasted it with what he called problematic inference, in which a conclusion is always less certain than its premisses (as typically in scientific inferences from experiments to generalisations). But this kind of terminology and classification does not illuminate the logic of the situation. It is based on psychological descriptions of the states of mind allegedly associated with certain types of inference, rather than on an analysis of the logical relations between the premisses and conclusions of those inferences. If we look instead to the latter we shall see how the ability of a single instance to give full support to certain generalisations is just a special case of the type of support that is commonly given by more complex experimental results, and how, *pace* Johnson, no greater certainty is justifiable in the former case than in the latter.

The point emerges quite clearly if we consider Johnson's own example.[1] 'In judging upon a single instance of the impressions red, orange and yellow', he said, 'that the qualitative difference between red and yellow is greater than that between red and orange (where abstraction from shape and size is already presupposed) this single instantial judgement is implicitly universal; in that what holds of the relation amongst red, orange and yellow for this single case, is seen to hold for all possible presentations of red, orange and yellow.' Maybe so. But if we wish to articulate the logical structure of the inference, rather than describe the mental experience of making it, we cannot ignore the fact that, as Johnson put it, 'abstraction from shape and size is already presupposed.' The evidential proposition reports both that some red impression is more unlike some yellow impression than it is unlike some orange impression, and also that this instance of a difference between degrees of unlikeness is unaffected by any variant of the other variables that are relevant to the comparison of impressions, viz. shape and size, as in an evidential proposition like (7). Correspondingly the hypothesis that is supported is a generalisation somewhat like (4). Admittedly, Johnson intended his term 'intuitive induction' to embrace certain inferences about non-empirical subject-matters, like those of ethics or mathematics, as well as certain inferences about sensory impressions, like the one in the above example. But it will be shown in the following two sections, §§17 and 18, that the logical syntax of these inferences is inductive in

[1] Op. cit., p. 192 f. Johnson is going back, as it were, on an insight of Hume's. Hume saw clearly that no special mode of reasoning is involved in this kind of case. Cf. *Treatise* bk. I, pt. III, sec viii, Selby-Bigge ed. p. 104.

precisely the same sense as the inference from (7) to (4) or from (8) to (1).[1]

A critic sympathetic to Popper's ideas may now be inclined to object that, if U′ is a version of U which has relatively low-grade simplicity, then U′ can earn full support too easily according to the method of assessment that has just been proposed for hypotheses like (1), (4) or (5). The relative complexity of U′ ought to count against it, he may think, not in its favour. But this is to make U′ suffer for the sins of U. If grade of inductive support is to depend on the answers Nature gives to the experimenters who interrogate her, a hypothesis should not be penalised for its complexity. For in some fields the truth about Nature may be rather complex, or rather complex hypotheses may be required in order to achieve the level of idealisation that is needed for the amalgamation of elementary hypotheses into correlational generalisations or for the unification of the latter into scientific theories. Where a certain complexity is desirable, the fault lies not with the version that has this complexity but with the version or versions that have not got it. Moreover we must be particularly careful here to distinguish between criteria for assessing acceptability and criteria for assessing support. It may well be that the only version of a hypothesis that can obtain a high grade of support is one that has too low a grade of simplicity to be acceptable as the answer to a certain kind of scientific question. But then the relative complexity of this version counts against its acceptability, not against its attaining a high grade of support. If simplicity is thus a merit in regard to acceptability, the assessment of this merit must be made in terms of a simplification-function, not a support-function. We want to know just how simple a certain hypothesis can be while still receiving full support from available evidence.

Up to a point, however, one can compare what is assessed by a simplification-function with what Popper called degree of corroboration. Popper's recommendation was to search for the simplest hypothesis that

[1] Conceivably one particular performance of test t_i, that results in full support for a generalisation like (4), may turn out to be the only occasion in the real world on which the antecedent and consequent of this idealised generalisation are co-instantiated (or approximately co-instantiated). On all other occasions on which the generalisation is invoked allowance may need to be made for various divergences from ideal conditions. Similarly one particular trial-outcome that verifies an evidential report like (8) may not only give full support to a generalisation like (1) but also be the only occasion on which this heavily qualified generalisation's antecedent and consequent are co-instantiated. The complex combination of circumstances that caused World War I, for example, may in fact never exactly recur, though if it did it would presumably cause a similar war. So the generalisations implicitly invoked by many causal explanations of unique historical events are in principle open to inductive justification, even though in practice such justification is often rendered very difficult by the numerousness of the variables instantiated in human life and the difficulty of contriving controlled experiments to discover which variables are relevant to which kinds of historical hypothesis.

escapes falsification by the severest available test, and a simplification-function does enable us to judge which of two hypotheses, in a particular field of enquiry, requires less complication if it is to avoid falsification. Indeed where two hypotheses like (2) are unequal in grade of permissible simplification, on given evidence, we shall often also have a corresponding inequality in degree of corroboration between the simplest fully supported versions of the two hypotheses, so far as such an inequality is determined by what Popper called the method of the sub-class relation.[1] But unfortunately Popper also held that absolute degree of corroboration, for a successfully tested hypothesis, was determined by its content, falsifiability or logical improbability. So, when he came to regard his concept of degree of corroboration as the measure of 'the intuitive idea of degree of support by empirical evidence',[2] he ended up with a proposal for a probabilistic measure of evidential support that ignored the empirical grounds on which relevant variables are selected for manipulation in experimental tests. If, on the other hand, we begin by recognising the crucial importance of this empirical basis for the construction of support-functions, we shall come to see that it is equally important in the assessment of simplicity for inductive purposes. We shall reject Popper's thesis that the simplicity of a hypothesis is to be thought of as a logical improbability, and accept that for inductive purposes grade of simplicity is no more a function of any logical probabilities than is grade of support.

§17. Inductive support for legal hypotheses

Inductive functors can take a non-standard interpretation in regard to permissible types of filler for their argument-places, as well as in regard to the type of grading (e.g. simplification instead of support) that they signify. For example, it can be shown that assessments of support for the conclusions of legal arguments from judicial precedent have the same logical syntax as assessments of experimental support for scientific hypotheses. The more obvious dis-analogies between legal and scientific reasoning stem from differences of subject-matter rather than of logic. The objection that legal argument issues typically in singular conclusions stems from pre-occupation with the role of judges, as distinct from academic lawyers; and a claim that it is analogical in character, or that it is often concerned to establish meaning rather than truth, is not at all incompatible with its being inductive. But the theory that legal argument from judicial precedent is inductive must be regarded as a

[1] *The Logic of Scientific Discovery*, p. 115 ff.

[2] *The Logic of Scientific Discovery*, p. 393. Cf. also the remarks on Popper's theory in §14 above.

deliberately idealised reconstruction, not a journalistic description. We ignore, for the moment, other types of legal premiss, occasional inconsistencies between precedents, the variety of forms in which actual legal rules are cast, and the controversies that have raged about the proper role of precedents in English, American or Australasian common law.

A non-standard role has now been established for inductive functions in regard to judgements about the relation between scientific hypotheses and reports of experimental test-results. Specifically, a function for assessing how much simplification is permitted by experimental evidence has been shown to be logically or analytically equivalent to the corresponding support-function and therefore subject to all the syntactic principles to which the latter is subject. But inductive functors can take a non-standard interpretation in regard to the permissible types of filler for their argument-places, as well as in regard to the type of grading they signify. Some inductive support-functions and simplification-functions may plausibly be supposed to be concerned with different types of subject-matter from those with which experimental science is characteristically concerned.

This is quite a crucial point. For if we wish to distinguish the syntax from the semantics of inductive functions there is a case for saying that we should include in the former every feature that is common to all inductive functions. It would be rather arbitrary to pick on features, like conformity to the general conjunction principle, that hinge on the occurrence of terms like 'and' within a hypothesis, and exclude features that hinge on other pervasive terminological characteristics. If all inductive functions were in fact concerned with the subject-matter of experimental science, despite the different branches of science to which they relate, then their syntax should perhaps be conceived to include whatever principles hinge on using the terminology of experiment and observation. That is, instead of just conceiving E, in a typical inductive assessment $s[H, E] \geqslant i/n$ to be a conjunction of one or more existential propositions, perhaps we should at the same time always conceive the truth of E to be experimentally ascertainable and work out the syntactic implications of the latter conception as well as of the former. Such a procedure would certainly appear to contaminate logic, as more narrowly conceived, with epistemology or heuristic methodology. But it can be shown to be wrong only so far as we can show that a quite substantial set of syntactic principles for inductive functions is unaffected by the difference between an experimental and a non-experimental subject-matter. For, as far as this can be shown, it becomes reasonable to draw the line between syntax and semantics in another place, and to treat the peculiarities of experimentalist terminology as a problem about the latter rather than about the former.

We cannot expect to find inductive functions in explicit use with regard to some non-experimental subject-matter, any more than we can claim them to be in explicit use within contemporary natural science. The concept of an inductive function is presented as an analytically idealised reconstruction, or explication, of actual patterns of reasoning—not as a true-to-life, warts-and-all representation of them. But it can be shown that such a reconstruction does fit certain patterns of legal reasoning from judicial precedent, just as well as it fits reasoning from the results of experimental tests, and that it also fits some other, analogous patterns of reasoning.

Let us consider elementary legal hypotheses of the form

(9) For any persons x and y, if x has R to y, then x has a good cause of action against y (i.e. if x sues y x ought to win),

where R denotes some legal or factual relation. For example, y might be the landlord of an inn at which x stayed for the night and y might have failed to stable x's horse in a way that preserved it from harm. Or, to take an example from criminal law instead, y might have killed a person in the territory over which x is sovereign. To establish such a proposition from judicial precedent it is necessary to show that in an appropriate variety of R-type situations x did indeed have a good cause of action against y. But what constitutes an appropriate variety of situations here?

In order to determine the appropriate variety of situations we have first to settle the branch of law, i.e. the class of materially similar hypotheses, to which our legal hypothesis belongs. A juristic support- or simplification-function must be assumed to be defined in terms of the relationships regulated by such a branch, e.g. murder, marriage, employment or innkeeper's hospitality. Next we have to determine the legal variables that are relevant to this branch of law. For example, the variables relevant to an innkeeper's hospitality might include such things as the size of the inn, the colour of a guest's skin, the length of his stay or the nature of his means of transport, and their relevance is determined analogously to the way in which the relevance of natural variables is determined. A certain hypothesis about an innkeeper's duty of care for a horse might be refuted by a case in which, say, the guest had been a casual daytime visitor, but it might apply in an otherwise similar case in which he had been an overnight resident. That is, in the latter case someone won his case under the rule suggested by the hypothesis, and in the former someone lost it. If the cases were otherwise similar, it is clear that the length of the guest's stay was a material fact in both cases, and hence that in general length of stay is a relevant variable in this branch of law.

It is easy enough to describe now the variety of evidence that should be required for a testable legal hypothesis like (9) to be fully supported by judicial precedent. We need records of x's having won the last round of a case against y before a court of appropriate jurisdiction, when x had R to y, in each possible combination of the variants of relevant variables, and in each case no other circumstances may be material than some such variant or combination of variants. No doubt it is rather unrealistic to suppose that such a systematic wealth of inductive evidence will ordinarily be available. But this is not an objection to the possibility in principle of inductive support-functions for legal hypotheses. For it is conceivable that a test-case could be brought in order to determine the outcome for each combination of mutually compatible variants. Indeed the possibility of so doing helps to fill out the analogy with experimental science. Moreover by turning our attention to less simple hypotheses than (9) we can draw inductive conclusions from very much less evidence. A single precedent, even, can suffice to give full support to an appropriately qualified hypothesis. Such a hypothesis would just need to be of 1st grade simplicity, concerning itself only with cases in which a certain combination of variants—one from each relevant variable—was present. Just as in experimental science the same cause is assumed always to produce the same effect, so too in legal reasoning it is assumed that like cases should be treated alike.

But what about the situation when there is less than full support? Any proposal to devise a measure for the relevance of natural variables encounters, as we have already seen (in §6), apparently insurmountable difficulties. Analogous difficulties must obviously block any attempt to devise a measure for the relevance of legal variables, and therewith any attempt to treat inductive support from judicial precedent as a measurable quantity. But in order to be able to fall back here on an assignment of ordinal, rather than cardinal, values to our support-function we need to be able to assume a cumulative hierarchy of varieties of evidence, like the cumulative hierarchy of tests that was assumed for hypotheses in experimental science. So we must postulate (as in §7) that somehow a well-ordering can always be determined for the set of variables that are relevant in a particular branch of law. The main way to do this is obviously to postulate that variables always have the same order in our support-theory as a competent judge would assign them an order of relevance, and that a competent judge looks to past cases to determine that order. If one variable has forced more rules to be qualified than another he will take it to be more relevant. That is, if a variant of variable

v_i is more often found in accepted versions of legal hypotheses in this field than a variant of v_j, then v_i is more relevant than v_j. For it is just this criterion of relevance that is most serviceable in establishing how much support exists for a hypothesis. It is obviously more important to discover precedents in each variant of v_i than in each variant of v_j, if the former variable has forced more rules to be qualified than the latter. Of course, safeguards are necessary, as in §7, against too frequent reversals in the supposed order of relevance, and a supplementary ordering relation will need to be invoked if two or more variables are supposed equally relevant. But there is clearly no difficulty in principle in constructing a cumulative hierarchy of varieties of evidence, i.e. of sets of precedent types, that would perform a role for legal support-functions quite analogous to that performable for experimental support-functions by a cumulative hierarchy of tests. Moreover the characteristic claim that must be made when extracting the actual precedent from any recorded judgement—the claim that no other circumstances were material than those mentioned in the alleged *ratio decidendi*—is precisely analogous to the claim made by $T_i x$ in (1), (2) and (3) of §7.

It follows that every syntactic principle that holds for experimental support-functions in virtue of this cumulative hierarchy must also hold for legal support-functions. The general conjunction principle, e.g., and the consequence principle for evidential propositions are both as valid for the latter as for the former. Also, it is very difficult to see how the consequence principle for hypotheses can fail to apply: how could the formal logical consequences of a legal rule be worse supported from precedent than the rule itself?[1] Hence the equivalence principle must also apply to legal support-functions, since both the consequence principle for evidential propositions and the consequence principle for hypotheses apply. Again, justice itself seems to require that the instantial comparability principle apply to first-order legal generalisation. If Smith has a better case in his circumstances than Brown has in his, the law would not be impartial between persons if there were not better support for anyone's having a good cause of action in Smith's circumstances than for anyone's having a good cause in Brown's; and if the latter proposition is true impartially then the former must also be true. So the instantial conjunction principle also applies to legal support-functions, at this level, since it is derivable, as in §2, from the equivalence and instantial com-

[1] Conceivably there are some situations in which, though I is a logical consequence of H, nevertheless rules of procedure prohibit a court from taking witnesses' testimony to the truth of H as being also testimony to the truth of I. But we are not concerned here at all with evidence about matters of fact, in the sense of witnesses' testimony or affidavits—only with evidence about matters of law, in the sense of judicial precedents.

parability principles. Two murderers are as rightly convicted as one. But, if there are any higher-order legal generalisations (see below), they are presumably not subject to any instantial comparability, or instantial conjunction, principles since these principles would produce antinomies here that are quite analogous to those they produce for theories and correlational generalisations in natural science (cf. §9).

In short, an appropriate axiomatisation of all these principles, as in §22 below, will serve to deploy the logical syntax of legal support-functions just as adequately as it deploys that of experimental support-functions. Such a syntax can now be clearly distinguished from the subject-matter of the functions concerned, and it becomes reasonable to define the concept of an inductive functor in terms of it.

However, jurists may be tempted, on various grounds, to dispute the account of legal reasoning from judicial precedent that has just been given.

First, they sometimes tell us that the outstanding difference between the deductive and inductive methods lies in the source of the major premiss: the deductive method assumes it, whereas the inductive sets out to discover it from particular instances. It follows, they argue, that judicial decision does not characteristically involve induction because it does not consist essentially in discovery of facts. An inductive scientist is seeking to discover and describe what is the case, a judge to prescribe what ought to be.[1]

But though an inductive scientist may well be seeking to describe what is in fact the case it does not follow that an inductive lawyer, if he exists, must be assumed to be doing the same. So far as we view his activity heuristically we must rather suppose that he is seeking to find out what ought to be the case—i.e. how the courts of a certain jurisdiction ought to decide the issues that come before them.

Nor are we confined to a heuristic conception of inductive inference any more than to a heuristic conception of deductive inference. The logical problems here are primarily concerned with criteria of justification, rather than with methods of discovery, in regard to both types of inference. They are concerned with what justifies conclusions, not with how the latter are reached.

[1] E.g. G. W. Paton, *Jurisprudence* (1964), p. 171 f. Analogies between legal and scientific reasoning have often been loosely stated, but seldom developed with any degree of systematic rigour. So lawyers may readily be excused the suspicions they often have about them. But it is odd that Bacon seems to say nothing about induction in his legal writings, and nothing about legal reasoning in his writings about induction. One would otherwise suppose that his training in common law methods helped to inspire his philosophy of induction. He was certainly prepared to grant that inductive reasoning could be applied to a non-empirical subject-matter, as by Plato in some of his discussions of definitions and ideas: cf. *Novum Organum* I, cv.

Perhaps it will be said that, even so, inductive inference is from proposition to proposition, and that no proposition can fill either argument-place in a legal support-functor, since legal reasoning is concerned with what ought to be, rather than with what is in fact the case. But it is important not to confuse the concept of a proposition with the concept of an empirical judgement. The concept of a proposition requires a logical definition, not an epistemological one. Now, a legal hypothesis like (9) may be assigned a truth-value in accordance with a Tarski-type truth-schema, since (9) may be said to be true if and only if for any persons x and y, if x has R to y, then x has a good cause of action against y; the usual truth-table definitions may then be set up for logical relations between such hypotheses; and the propositional calculus may thus be interpreted as a theory of deducibility-relations between legal propositions in a way that is quite on all fours with whatever can be done to interpet it as a theory of deducibility-relations between scientific propositions. If, therefore, legal hypotheses can safely be treated as propositions for the purposes of deductive logic, there seems no good reason to deny that they can be so treated in inductive logic.

Secondly, it is often claimed, as a further feature of disanalogy connected with the empirical basis of natural science, that any parallelism between legal reasoning and the reasoning of an inductive scientist is bound to break down in respect of whether or not the actual outcome of an individual case or experiment is relevant to the general rule under which it falls. The scientist, it is said,[1] who supports his prediction that some as yet unobserved R will be S from evidence about past R's having been S, gains further support for the rule legitimating this prediction, if the prediction itself turns out to be correct. But, it is objected, the judge who supports his own decision by reference to the *ratio decidendi* of previous cases cannot, without circularity, argue that his present decision gives further support to the correctness of his views about the relevant rule. Nor, on the other hand, is the losing advocate's argument refuted by the adverse decision: after all, the decision might be wrong.

This objection makes the mistake of treating a feature of disanalogy that in fact arises from difference of subject-matter as if it arose from difference of logical structure. The subject-matter of legal reasoning is a system of rules, not of empirical facts. A hypothesis like (9) states, in effect a rule for deciding lawsuits.[2] Hence there are no predictions at

[1] E.g. A. G. Guest, 'Logic in the Law', in *Oxford Essays in Jurisprudence*, ed. A. G. Guest (1961), p. 189, and R. Cross, *Precedent in English Law* (1961), p. 206.

[2] It is of course also possible, and useful for some purposes, to regard a legal system as a set of judicially enforceable rules for the conduct of anyone under its jurisdiction. This is certainly how the laws of his country must appear to the ordinary citizen. But for the analysis of inductive reasoning from judicial precedent it is convenient to adopt a more

issue here, to be confirmed or disconfirmed by the occurrence or non-occurrence, respectively, of what they predict. The judge's action in awarding judgement is justified, not predicted, by a proposition like '*a* has a good cause of action against *b*', which may be detached by *modus ponens* from a substitution-instance of a proposition like (9) if witnesses' testimony, and the legal interpretation appropriate to this testimony, so authorise.

Indeed, it is misleading to compare the role of a judge or tribunal here with that of a scientist, as jurists generally do.[1] An academic laywer is like a scientist, insofar as neither's conclusions need affect his actions. But a judge functions rather as an engineer does. He has to make both an intellectual decision about which party has the better case—a decision to accept or reject the consequent of a substitution-instance of a hypothesis like (9)—and also a practical decision about whether to award judgement for that party. The two decisions are easily confused by philosophers because precisely the same string of words may express both the intellectual decision and also the declaration of award. But a dishonest judge makes the two decisions differently, and correspondingly the proposition, or intellectual decision, '*a* has a good cause of action against *b*' may be said to justify awarding judgement for *a* against *b*. Now, the proposition 'Any bridge built according to specification R is stable' may equally be said to justify an engineer's action in building such a bridge. However the justification here is based on a prediction about what will happen rather than on an evaluation of what ought to be done. Consequently the engineer has an opportunity to be borne out by what happens while the judge has not. But this difference stems from subject-matter, not logic—from being concerned with empirical facts rather than with rules.

Moreover, inductive support is timeless support for timeless propositions, not temporally relative support for time-bound people or their judgements. It is true that the engineer may be borne out by what happens while the judge can never be. But the acceptability of a legal proposition, on certain evidence, may vary over time (cf. §1 and §10), even though the inductive support cannot. A judgement for the plaintiff on Wednesday does not make the plaintiff's case more acceptable on Tuesday, though Wednesday's judgement may make future cases of this type more acceptable. Similarly the bridge's stability after it is built does not help to make the engineer's prediction of stability any more acceptable before it is built, though it may make similar predictions more acceptable

Kelsenian paradigm, like (9). The two interpretations do not conflict with one another. They are mutually inter-translatable reconstructions of actual discourse about law.

[1] E.g. R. Cross, loc. cit.

afterwards. But the grade of support given by such—legal or experimental—evidence to a universal hypothesis and to its substitution-instances is just the same, whether this support is assessed before the evidence comes into existence or afterwards. The most that we can say here is that once judgement has been awarded in the legal case another piece of potential evidence exists for the universal hypothesis, like (9), that constitutes its *ratio decidendi*, just as once the bridge has been safely built another piece of potential evidence exists for the universal hypothesis from which its stability was predictable.

A third kind of objection is sometimes raised against an inductive conception of legal reasoning from judicial precedent. The logical form of inductive reasoning, according to many writers on law,[1] is: 'x and y repeatedly occur together, therefore if x occurs y occurs.' Even where variation of circumstance is considered important for induction, it is usually assumed that no worthwhile inductive support is forthcoming except from a multiplicity of instances.[2] But it follows from that assumption, runs the objection, that the characteristic method of legal reasoning from judicial precedent cannot be inductive, since a single binding precedent frequently suffices to establish the validity of a decision beyond any dispute.

Indeed, at this point some writers on legal reasoning from precedents seem to give up any attempt to characterise its logic further. Since such reasoning is neither deductive nor inductive, they seem to think, it is just *sui generis*, and nothing general can be said about its logic.[3] Other jurists do not give up quite so soon. Argument from judicial precedent is argument by analogy, they claim, and they suppose the difference between this form of argument, on the one side, and induction or deduction, on the other, to have been shown by Aristotle.[4] As a typical instance Aristotle gave the argument that it is evil for the Athenians to fight against the Thebans because it was evil for the Thebans to fight against the Phocians and the cases are analogous in that the Athenians are neighbours of the Thebans and the Thebans are neighbours of the Phocians.[5] In Aristotle's view this kind of reasoning was from part to part, where the relevant whole is some general principle such as that it is evil to fight against neighbours. He therefore distinguished it both from inductive reasoning, which he regarded as being from parts to whole, and

[1] E.g. A. G. Guest, op. cit., p. 188.
[2] E.g. O. C. Jensen, *The Nature of Legal Argument* (1957), p. 28.
[3] E.g. O. C. Jensen, op. cit., pp. 7–31.
[4] E.g. E. H. Levi, 'An Introduction to Legal Reasoning' 15 *Univ. Chicago L.R.* (1948), p. 501 ff. Cf. R. Cross, op. cit., p. 207 ff., and R. Stone, 'Ratiocination not Rationalisation', *Mind* lxxiv (1965), p. 463 ff.
[5] *An. Pri.* II, 24.

from all other kinds of reasoning, which he regarded as being normally from whole to part.

But such an objection is based on false premisses. We have already seen (§16, p. 152 f.) that there is no difficulty whatever in supposing a sufficiently complex version of a universal hypothesis to be fully supported by a single instance, whether in science or in law. Given the truth of the evidential report, a man is entitled to take the version to be as well supported as his belief that just such-and-such variables are relevant to hypotheses of this kind, which is the belief that underlies all his other assessments of inductive support in the same field. Perhaps confusion may be created here by the old classification of so-called enumerative induction alongside induction by variation of circumstances as two species of the same logical genus. This classification may suggest the view that induction by variation of circumstances is, as it were, a more discriminating brand of enumerative induction—retaining the latter's respect for multiplicity of instances and adding its own respects for instantial variety. On such a view no generalisation over an unbounded domain could ever be fully established by a single instance. But if enumerative and variational induction are not species of a single logical genus, because the syntax of the former is quite different from the syntax of the latter (cf. §§12–14), we are quite free to treat establishment by a single instance as a special case of the latter type, where the proposition supported has 1st grade simplicity.

We shall no doubt now be reminded that the jurists' case against induction does not rest solely on the frequent appeal to a single, singular proposition as sufficient premiss for a legal conclusion. This conclusion too, it may be said, is characteristically a single, singular proposition, whereas the typical conclusion of an inductive argument is a universal proposition. Has not Aristotle shown that such reasoning, from singular premiss to singular conclusion, is analogical, not inductive?

But Aristotle has only shown that such reasoning is not inductive in the sense in which he himself was using that term,[1] viz. to describe a method of reasoning that proceeds through an enumeration of all the cases. If instead we understand the term in Bacon's sense, it is equally clear that analogical argument is, essentially, inductive. Indeed, the thesis that analogical inference is intrinsically different from induction was taken up from Kant by Sir William Hamilton and long ago trounced by J. S. Mill.[2] Aristotle's authority does not justify its revival. We may accept that it was evil for the Athenians to fight against the Thebans because (i) it was evil for the Thebans to fight against the

[1] *An. Pri.* II, 23. Contrast *An. Post.* I, 1.

[2] *An Examination of Sir William Hamilton's Philosophy* (1865), p. 402 f.

Phocians and (ii) the cases are analogous in that the Athenians are neighbours of the Thebans and the Thebans are neighbours of the Phocians. But this argument is only valid so far as it was just their being neighbours that made it evil for the Thebans to fight against the Phocians and also no relevant circumstance differentiates the Athenian-Theban situation from the Theban-Phocian situation. So the premiss of the argument is a proposition like (7) and the conclusion is a substitution-instance of a generalisation like (4). Moreover any other substitution-instance of the same generalisation could equally well figure as the conclusion. Hence it is clear that Aristotle's so-called analogical argument is just a species of inductive reasoning. It derives from the uniformity principle—cf. (23) of §4—that a first-order universal hypothesis and each of its substitution-instances are equally supported by the same evidence. If (7) supports (4) it supports also each substitution-instance of (4).

In any case it is wrong to hold that legal argument from judicial precedent issues exclusively in singular conclusions. It is true that a judge, who has to decide the single case before him at any time, does not need to draw other than singular conclusions, even if he sometimes does do so. He is, as has already been remarked, like an engineer who has to design a single bridge of predictable stability. But an academic lawyer, or text-book writer, is often concerned to establish general hypotheses that state rules of law, and may even seek to unify certain sets of hypotheses into a relatively comprehensive theory, as a theoretical scientist does. Carnap's unduly narrow conception of scientific reasoning as engineers' reasoning (cf. §14, p. 130 f. above) has its counterpart in the unduly narrow conception of legal reasoning from precedent as judicial reasoning by analogy.

Fourthly, it is sometimes said,[1] in connection with the structure of legal reasoning, that a great deal of it is concerned to establish the correct meanings of legal rules rather than their truth or validity, and that this preoccupation with the determination of vagueness is a feature of legal reasoning which sharply differentiates it from any kind of deduction or induction. But in fact, as we have seen in the preceding section (§16), for every inductive support-function there is an equivalent simplification-function that is concerned with just this issue. When a line of precedents has exhibited a series of eroding distinctions, a less and less simple version of a certain legal hypothesis has become tenable, and correspondingly the appropriate monadic simplification-function is one that has to be assigned lower and lower values for this hypothesis. Later judges have decided that certain circumstances were material in earlier cases, which

[1] E.g. Jensen, op. cit., p. 28, and Levi, op. cit., p. 502.

may not actually have been thought so at the time. For example, the casual daytime visitor to an inn, whose horse need not be stabled by the innkeeper, may be distinguished from the visitor whose horse had to be stabled because he was an overnight resident. Overnight residence is now seen to be a material factor in the earlier case. Or, in other words, it is like the discovery of a hidden variable in experimental science. Length of stay has now come to be regarded as a relevant variable of which one variant falsifies certain versions of the hypothesis, and any conclusion to be drawn from—any version to be fully supported by—the precedents must be suitably qualified in relation to this variable. In no case, perhaps, has a precedent been set aside, but in each of the earlier cases, the *ratio decidendi* has been reinterpreted. On the other hand, when a line of precedents extends a rule of law, the opposite process takes place. Evidence piles up so as to show that initial versions of the hypothesis were unnecessarily restricted: a simpler version has full support. For example, perhaps what was material to the innkeeper's duty to stable the visitor's horse was not the fact that it was an animal needing shelter, but the fact that it was the visitor's personal means of transport. Or, in other words, the *nature* of a visitor's personal means of transport is no longer regarded as a relevant variable of which some variants falsify the simplest version of the hypothesis. So there is as much support for the conclusion that an innkeeper has a duty to garage a visitor's motor-car as for the conclusion that he has a duty to stable a visitor's horse.Hence precedent-based arguments about the meaning or scope of an acknowledged legal rule are as much entitled to be called 'inductive' as are precedent-based arguments that are designed to justify acknowledging such a rule. The former are representable by inductive simplification-functions, the latter by inductive support-functions. By giving the concept of an inductive function a syntactic rather than a semantic definition, so that simplification-functions are no less inductive than support-functions, we contrive to bring out the underlying logical unity of the two main modes of legal reasoning from judicial precedent, which are superficially rather different from one another.

A fifth kind of objection to the possibility of legal induction is also worth considering. The inductive conception of legal reasoning, it may be objected, does not correspond with the untidy facts of common law jurisprudence. It is a product of philosophical imagination, we may be told, and misrepresents the ways in which English, North American or Australasian lawyers actually reason.

Even if this were wholly true it would not matter for present purposes, since our discussion has been designed to show only the possibility, not the actuality, of inductive argument from judicial precedent. It is only this that needs to be shown in order to establish the strength of the syntactic principles that have been proposed in earlier chapters—i.e. the fact that they are not intrinsically and necessarily confined, in their application, to an experimental subject-matter. But in fact the situation in regard to induction in common law is not much different from that of induction in experimental science. In both cases what is offered here is a logically idealised reconstruction of actual reasoning, not a psychological, historical or journalistic description of it. Just as in science itself no comprehensive theory can normally emerge in a given field without a certain element of deliberate idealisation—without, that is, setting certain deviant phenomena on one side, to be otherwise explained—so too in inductive logic our theory can only achieve power and interest if it is prepared to operate at a sufficient level of analytical abstraction (cf. p. 4). We can now go on to mention a few of the phenomena not covered by our theory. But these phenomena no more constitute an argument against the theory than the width of a visible line constitutes an argument against a Euclidean theory of space.

First, our account of legal reasoning from judicial precedent has said nothing about the way in which such reasoning is in practice often interwoven with appeals to other kinds of premisses, such as statutes, general maxims or principles of law, considerations of public policy, minority opinions, or *obiter dicta*.

But there are analogies for all, or nearly all, these other factors in experimental reasoning. A lawyer's appeal to statute or general principle is like an experimenter's appeal to accepted methodological precepts, mathematical theorems or linguistic definitions. For example, the consequence principle for hypotheses may be taken to imply that if H is a proposition of law and I is a consequence of H according to a statute in force, then I is at least as well supported by any statement of precedents as H. If a statute declares the latest date for filing tax-returns to be May 1, and x has a good cause of action against his accountant y if y fails to file x's tax-return by the due date, then x has a good cause of action against y if y does not file x's tax-return on or before May 1. Again, considerations of public policy are rather like the expectation of profit from relying on a scientific hypothesis if it is true, or of loss if it is false. Just as the latter expectation may affect the acceptability of a hypothesis, or of using such-or-such tests on hypotheses of a certain kind, so too considerations of public policy may affect the (precedent-creating) decisions a judge should make or the circumstances he should consider material in certain types of case. Also, the occasionally persuasive force of minority opinions, *obiter dicta*, decisions in other countries, etc. is like an experimenter's reliance on untested, but plausible or generally accepted assumptions, or on the opinions of Nobel prize-winners.

Secondly, there are a number of factors that in practice militate against consistency within a system of legal reasoning from judicial precedent. As the volume of decided cases accumulates it becomes more and more difficult for anyone to take all the cases into account that he should, so that inconsistencies may arise between the precedents set by judges who have taken one set of cases into account and the precedents set by those who have taken another set into account. Inconsistencies may also arise because of the size of a country and the varying importance attached to different interests in different regions, or the varying background and training of the judges involved.

But the existence of such inconsistencies in practice is no reason to suppose that an analytical reconstruction of legal reasoning should not assume an ideal of consistency. There have been long periods in the history of natural science in which mutually inconsistent experimental results have existed side-by-side in the texture of accepted belief. Notoriously, for example, a wave theory of light has been accepted on account of some results at the same time as a particle theory was accepted on account of others. So that on the one hand it is possible to point to the existence of inconsistencies in both natural science and the common law, which act as a kind of Hegelian yeast in the causal dialectic of their development; and on the other hand it is possible to claim that they would not act as a yeast if the ideal of consistency were not also present.[1] When therefore we analyse the concept of inductive support in abstraction from the history of its use, we are entitled to ignore the occasional occurrence of apparent inconsistencies between different bits of evidence, whether in law or in science. We may suppose that on most occasions, whether in law or in science, the appearance of inconsistency is merely superficial and will disappear when the operation of a hidden variable is revealed.

Admittedly, in law it is also possible that a precedent case was wrongly decided; and there is no systematic analogy for this in the subject-matter of experimental reasoning, unless we adopt a kind of Manichaean metaphysics, attributing some events to the action of the Deity and the rest to the action of the Devil. But the disanalogy here is just due to the fact that in law our subject-matter is a set of rules and in

[1] Curiously, an English judge (Lord Porter in Best v. Samuel Fox & Co., A.C. 727, 1952) has declared that 'the common law is a historical development rather than a logical whole and the fact that a particular doctrine does not logically accord with another or others is no ground for its rejection'. But to accept such a dictum is implicitly to allow that in common law it may be legitimate to treat like cases dissimilarly. Contrapositively, the requirements of natural justice impose an ideal of logical consistency.

science a set of facts. In relation to legal hypotheses like (9) we are dealing with a set of rules for judicial decisions. Within such a framework of rules both correct and incorrect decisions can occur, and therefore from a court's having decided that x had, or did not have, a good cause of action against y in a certain situation it does not necessarily follow that x actually had a good cause (unless the court is one that has irreversible precedent-creating jurisdiction). It is not the historical narrative that such-or-such courts made such-or-such decisions which is to be taken as a typical evidential proposition (reporting canonical test-results) for a legal hypothesis like (9), but rather certain legal propositions that may be read off from the narrative as interpretations of those decisions. On some occasions these interpretations may even be barred by rules of the legal system, e.g., by statute, or when the decision is subsequently reversed.

Thirdly, the rules of common law are of many different types, and by no means all of them are easily recognised as constituting hypotheses like (9). When one person has a good cause of action against another, it is normally because the defendant has not done something he ought to have done or has done something he ought not to have done. But it is a mistake, some jurists will claim, to suppose that all legal rules relate to what people ought or ought not to do. Many of them relate instead to what they may do, under the protection of the law, rather than to what they ought or ought not to do: many rules of law may be viewed as conferring powers on people, rather than as imposing duties on them. Hence it is a mistake to suppose that all reasoning from judicial precedents can be conducted in terms of hypotheses like (9).

It has been shown elsewhere, however, that H. L. A. Hart's dichotomy between 'duty-imposing' and 'power-conferring' rules is not one of any special importance.[1] There are certainly very many different types of legal proposition, just as there are very many different types of scientific hypothesis. But an analytical reconstruction of inductive reasoning about any subject-matter is bound to concentrate on certain paradigmatic forms, and to claim that, for every proposition about the subject-matter, there is an equivalent that has one or other of these forms. It is as unreasonable to object to this procedure in jurisprudence as in the philosophy of science. For example, a typical 'power-conferring' rule, such as one that allows people to marry, may be redescribed in terms of a set of hypotheses like (9) which ordain an outcome for each kind of case that might come to court about a desired, attempted, or existing marriage. It may well be that some of these legal rules or princi-

[1] Cf. L. Jonathan Cohen, critical notice of H. L. A. Hart's *The Concept of Law*, in *Mind* lxxi (1962), p. 395 ff.

ples, though non-quantitative in form, have a comprehensiveness or generality that makes statements of them more suited to comparison with the correlational generalisations, or theories, of natural science than with its relatively elementary hypotheses. For example, perhaps one such principle might be that any rule restricting the liberty of the normal citizen should be narrowly construed, or that the validity of a marriage is normally to be determined by reference to the law of the country in which it was contracted. Perhaps too a certain gift for conceptual innovation has sometimes helped the great systematising judges, like Lord Mansfield, just as it has helped the great systematising scientists. But none of this is a reason to reject the view that arguments from judicial precedent assume an underlying relation of inductive support.

An apparent difficulty is created by the American practice of prospective overruling,[1] whereby a court declares, quite independently of a legislature, that as from a certain date the law about such-and-such is to be so-and-so. It may appear that the legal hypothesis stated by a declaration of this nature cannot be a universal proposition, since it is temporally restricted, and that therefore it cannot be a subject of inductive reasoning. But in fact such a hypothesis can be regarded as a generalisation over an unbounded domain, since it purports to affect *all* actions after a certain date. The difficulty is not so much the logical one of regarding the hypothesis as a universal proposition, but rather the legal one of accepting that there is any inductive evidence that could be offered for the restriction of date—just as it is difficult to accept that there is any evidence (canonical test-results) for saying that all emeralds are grue, in Goodman's sense (cf. §11 above), though the hypothesis that they are all grue is respectably testable in form. The American courts that have engaged in prospective overruling have promoted legislation in the form that is more characteristic of a legislature than a court, viz. to be valid only after a certain date, and such justification as may exist for these rulings seems correspondingly to be more political than legal in nature.

Fourthly, a theory that asserts argument from judicial precedent to be isomorphic with argument from scientific experiment may be accused of taking sides with one or other of a number of highly controversial juristic doctrines that have sometimes even influenced the actual development of certain legal systems. Such a theory may be accused, for example, of taking sides with Blackstone's doctrine that judicial precedents declare pre-existing customary law as against Austin's doctrine that

[1] For the introduction of the doctrine of prospective overruling into Indian jurisprudence cf. O. Hood Phillips, 'Fundamental Rights and Prospective Overruling in India', *Law Quarterly Review* lxxxiv (1968), p. 173 ff.

legal induction is a process 'through which a rule made by judicial legislation, is gathered from the decision or decisions whereby it was established.'[1] But a theory about the logical syntax of inductive support-functions is inevitably neutral between Blackstonian realism and Austinian conceptualism, since it is concerned with a timeless, logical relation between propositions. It implies nothing about the temporal priority of rules to decisions or of decisions to rules, just as it implies nothing (cf. §1 above) about the temporal priority of hypotheses to experiments or of experiments to hypotheses. Again, the theory may be accused of taking sides with the doctrine prevalent among twentieth-century English lawyers that precedents (except perhaps in the House of Lords) normally have binding force, as against American doctrines that qualify the force of precedent by a judge's right to ignore precedents which conflict in one way or another with current social or economic circumstances. But in fact the theory is quite neutral between such opposing legal doctrines.[2] It is compatible with the existence of any definite number and variety of mutually consistent constraints on the title of a judicial precedent to constitute a piece of inductive evidence. These constraints affect the semantics of legal support-functions, not their syntax.

Our theory of legal induction, however, is certainly incompatible with a strictly realist, Holmes-like analysis of statements about common law, since it is intended as a reconstruction of the reasoning with which not only advocates and academic lawyers but also judges themselves can support their conclusions about the law. Judges can hardly take these conclusions, as Holmes seems to have suggested, to be predictions about what they themselves *will* do, since, as we have seen, the legal conclusions concerned constitute reasons why judgement *should* be awarded to one party rather than the other. These conclusions must take the form of saying that because of certain circumstances x has a good cause of action against y, rather than of saying that because of certain circumstances x will turn out to win his case against y. But if a realist, Holmes-like philosopher prefers the latter form of analysis for typical statements about common law (and, no doubt, some legal advice is actually given in this form) the mode of legal argument he envisages must still, at least in part, be inductive. It will just be about facts, as in natural science, instead of rules, and may accordingly combine statistical modes of reasoning with inductive ones.

[1] John Austin, *Lectures on Jurisprudence*, 3rd ed. (1869), p. 66.

[2] Lawyers' insistence on the binding force of precedents, within a common law jurisdiction, tends to vary with the ease of changing certain features of the law by the action of the legislature. It was perhaps narrower and weaker in eighteenth-century England than now, and wider and stronger in early twentieth-century U.S.A. than now.

§18. Universalisability, evaluation and semi-grammaticalness

It is important to distinguish syntactic from semantic elements in principles of natural uniformity, ethical universalisability, legal impartiality, etc. Also, in addition to inductive support-functions for legal, ethical or grammatical hypotheses we can construct inductive evaluation-functions for legal merit, moral propriety or linguistic grammaticalness. The concept of an inductive evaluation-function for grammaticalness may be used to elucidate the problem of how to describe a native-speaker's ability to comprehend semi-grammatical sentences or a child's ability to produce them; and the applicability of inductive functions to reasoning in both fields points to certain interesting analogies between legal systems and systems of grammatical syntax.

It is worth noting how certain principles of inductive syntax turn out closely similar to principles that are already familiar under other names. Consider, in particular, the uniformity principle for dyadic functions, (23) of §4, viz. where P is a substitution-instance of a first-order universal hypothesis U, and E is any normal statement of inductive evidence,

$$s[P, E] = s[U, E].$$

Since one of the cases in which this will hold is when E is true and reports every piece of evidence for or against U that ever exists, we also have a uniformity principle for monadic functions, viz.

(10) $s[P] = s[U]$.

And since one of the cases in which (10) will hold is when $s[P] = n/n$, (10) implies that any inductively establishable singular truth is an instance of an inductively establishable general truth. But what does this mean in terms of the experimental and common law interpretations of the monadic inductive functor? On the experimental interpretation (10) implies the uniformity of nature, so far as nature is inductively knowable, and in the particular case of causal hypotheses it implies the validity, so far as we can ever know, of the familiar principle: same cause, same effect. On the common law interpretation (10) states a familiar principle of natural justice for the special case of precedent-based decisions: like cases should be treated alike.

Note that nothing is stated by (10) about U's domain of discourse or the meanings of its terms, so that (10) is quite compatible with U's being statistical in character, like (2) of §13, or with U's stating a rule of law that is only valid after a certain date. If anyone wishes to assert that the world of nature is fundamentally non-statistical in character, his thesis is scientific or metaphysical, not logical, and within the idealised natural-scientific semantics for (10) he must impose a further restriction on the possible meanings for U. Similarly if anyone wishes to assert that all rules of common law have a scope of application that antedates their first explicit declaration his thesis is legal or juristic, not logical, and he must impose a corresponding restriction within the idealised legal semantics for (10).

This difference between the apparent catholicity of (10) as a purely logical principle, and its capacity for parochialism when semantically reinforced, is especially familiar in the principle of universalisability—or the golden rule—in ethics. If it is (normally) wrong for you, *qua* a human being, to be enslaved, I may say, it is wrong for anyone to be enslaved. But so far as logic goes, you are equally entitled to say instead that if it is wrong for you, *qua* a Greek, to be enslaved, it is wrong for any Greek to be enslaved, and then by accepting the antecedent of this conditional, rather than by accepting the other, you commit yourself to a good deal less—which may be convenient if you happen to be a Greek. The only premiss that, according to the principle of universalisability, must lead you to the conclusion that it is wrong for anyone to be enslaved is the premiss that it is wrong for you to be enslaved irrespective of your special circumstances. Of course, you may well think that nationality is not a relevant variable at all for elementary moral generalisations like

(11) For any x and y if x enslaves y, x acts in a morally wrong way towards y,

and therefore that no inductive version of (11) can have its antecedent complicated by such a qualification as 'and y is a Greek'. You will then hold that there cannot be any greater support for the wrongness of enslaving Greeks than for the wrongness of enslaving human beings in general. But it is your conscience, or moral sensitivity, not your grasp of logical syntax, which tells you that nationality is not a relevant variable for moral hypotheses like (11).

So the old debate about whether the principle of universalisability is morally neutral or not may be easily resolved. As a principle of logical syntax, asserted by (10), universalisability is implicit in any inductive reasoning about moral issues and is accordingly quite neutral. But it then tells us nothing about the proper place of nationality, religion, colour, etc. in our moral judgements. It is compatible with any ethic, however egotistical. As an oecumenical thesis, on the other hand, which asserts that moral rights and duties are unaffected by nationality, religion, colour, etc. universalisability presupposes that certain important restrictions have been accepted within the ethical semantics for inductive functors. Certain variables have been accepted as being irrelevant. In this form the principle of universalisability is no longer neutral, and its non-neutrality is shown up by the fact that it is no longer just a principle of logical syntax, like (10).

We have been assuming in the last two paragraphs that, when moral generalisations are supported by the dictates of conscience in individual cases, the support is inductive in its logical syntax. It would be unnecessarily tedious to argue this in detail. But in an idealised reconstruction we should have to suppose a Tarski-type definition for the truth of moral propositions,[1] as for legal ones, and an office for conscience, or moral sensitivity, that is analogous to the function of judicial precedent or scientific experiment. Also we should have to suppose a possible plurality of ethical support-functions. One reason for this is that there may be different fields of moral enquiry, as of legal or scientific reasoning: e.g. personal, commercial, professional morality, and so on. Another reason is that there may be irreducibly different forms of moral proposition: e.g. propositions about supererogatory virtues as well as propositions about right and wrong. Yet another reason is that there are some counterparts in ethics to the attempts of scientists or lawyers to organise some of their hypotheses into systematic theories or doctrines. Some concepts of virtues, like justice, courage or kindness, seem to embrace a whole field of generalisations about how it is proper to behave, and any proposition about such virtues (e.g. 'Kindness is more important than courage') may therefore have indefinitely many implications of a more elementary type. Moreover a theory about virtues in general, like Aristotle's doctrine of the mean, or a unifying first principle, as in Mill's utilitarianism, operates at a yet higher level of generalisation. Similarly moral generalisations, like legal or scientific ones, are often adjusted in scope or meaning, rather than merely accepted or rejected. Indeed, first-order moral generalisations, e.g. the rule not to tell lies, tend to be formulated, like familiar causal generalisations, in a way that makes them readily applicable to normal circumstances, even though they are inapplicable to certain abnormal ones; and Aristotle's doctrine of the mean, as a higher-order theory, had to be formulated over a domain of idealised entities—the brave man, the temperate man, etc.—just like correlational generalisations in natural science. So a corresponding plurality of simplification-functions would also be needed.

Even the same difficulties recur. For example, someone may object to the conjunction principle for moral hypotheses that, though there is no

[1] Just as sentences like (11) may therefore be treated, for logical purposes, as propositions, so too expressions like 'acts properly towards' may be treated, for logical purposes, as predicates. No distinction between science and ethics can be drawn at the level of logical *syntax*. But we do not destroy the so-called 'is-ought' distinction in this way, since it appears now as a *semantical* distinction between different types of predicate.

support for its being wrong to prevent someone who is depressed from drinking several tots of whisky and also none for its being wrong to prevent someone who cannot sleep from taking a barbiturate, there is nevertheless a great deal of support for its being wrong not to prevent someone who is depressed and sleepless from taking both whisky and barbiturate. But this object is answered quite analogously to the one that was discussed in §8 above. A highly relevant variable, in relation to what it is proper to give people, is what they have already been, or are concurrently being, given. So a hypothesis about the permissibility of supplying whisky to someone who is depressed, or barbiturates to some-one who is sleepless, must be suitably qualified in its formulation or suitably restricted in its domain of discourse if it is to have any substantial support, and then the conjunction principle creates no paradox.

It is noteworthy too that legal and moral reasoning both allow another type of interpretation—or rather, a partial interpretation—for the inductive functor, besides interpretation of it as the expression of support- and simplification-functions. Whereas the latter functions grade or rank the hypothesis itself, this other type of function grades the wrongness asserted by the evaluative consequent of a moral hypothesis like (11), or the merit of the case asserted by the consequent of a legal hypothesis like (9), in the light of what is assumed by its antecedent. Such evaluation-functions can also be either monadic or dyadic. For example, if U is a first-order moral generalisation, as in (11), we can read the assignment of an ordinal value to the monadic evaluation-function for U, viz. 'e[U] $\gtreqless i/n$, as 'For any persons x and y, if the only morally relevant feature of their relation is that x enslaves y, then x acts towards y with just (at least, at most) ith grade wrongness', where this has precisely the same sense as 'There is just (at least, at most) ith grade support for the best supported hypothesis that implies, for any persons x and y, the moral wrongness of x's enslaving y'. Similarly, if U is a first-order legal generalisation like (9), we can read a dyadic assignment 'e[U, E] $= i/n$' as 'For any persons x and y, if the only legally relevant feature of their relation is that x has R to y, then x has a cause of action against y that has ith grade merit on the (legal) evidence of E', where this has precisely the same sense as 'On the evidence of E there is just ith grade support for the best supported hypothesis that implies for any persons x and y, the merit of x's case against y if x has R to y'. Corresponding meanings can be given to 'e[P] $\gtreqless i/n$' and 'e[P, E] $\gtreqless i/n$', where P is a substitution-instance of such a U or an existential generalisation of such a substitution-instance.[1] Also, if H and H′ are both legal generalisa-

[1] Such a function enables us to make comparative judgements that are not restricted to comparisons between obligations, on the one side, and morally indifferent or prohibited

tions like (9) or moral ones like (11), or substitution-instances of these, or existential generalisations of substitution-instances, then '$e[H \,\&\, H'] < i/n$' may be given the meaning of 'Either $e[H] < i/n$ or $e[H'] < i/n$', and '$e[H \,\&\, H'] > i/n$' may be given the meaning of '$e[H] > i/n$ and $e[H'] > i/n$', and '$e[H \,\&\, H'] = i/n$' may be given the meaning of 'Neither $e[H \,\&\, H'] < i/n$ nor $e[H \,\&\, H'] > i/n$'; and similarly for the dyadic functor. But no analogous meanings seem available for evaluation-functors in other cases. Moreover any attempt to develop a function of this type for certain kinds of scientific propositions, as distinct from moral or legal ones, is rendered fruitless by the fact that there are normally other, and more useful, ways of grading the natural properties or characteristics designated by the consequents of scientific hypotheses (e.g. by weight, size, wavelength, etc.).

The idea behind such an evaluation-function is that as a legal hypothesis, like (9), or a moral hypothesis, like (11), becomes better supported, the version of it that is fully supported becomes simpler, i.e. less restricted; and the less restricted in scope is an established legal or moral generalisation, the more importance we attach to it, and *vice versa*. The wrongness of killing, for instance, may be thought more important than the wrongness of telling a lie, insofar as all circumstances that are exceptional for the former (war, self-defence, etc.) are also exceptional for the latter, while many exceptional circumstances for the latter (arising out of politeness, kindness, etc.) are not exceptional for the former. Hence, if the wrongness of doing a certain act has just ith grade importance, we may suppose that this is because an ith grade version of the best supported hypothesis to imply this wrongness is the simplest fully supported version of that hypothesis, and thus, in virtue of the equivalence between simplification- and support-functions that was pointed out in §16, the best hypothesis to imply this wrongness must have just ith grade support. No doubt there are other ways of grading the importance of a legal or moral generalisation than by reference to its width of scope, where this width is assessed in terms of a list of inductively relevant variables. For example, it is sometimes possible to measure or compare the utilities involved. But it is not uncommon to grade moral or legal importance in roughly the way described, and when anything is so graded the appropriate evaluation-functions will obey all the applicable principles of inductive syntax that have already been discussed. The logical syntax of such functions is a proper part of the logical syntax of inductive functions.

courses of action, on the other, or between morally indifferent actions, on the one side, and prohibited ones, on the other: contrast L. Aqvist's proposal in 'Deontic Logic Based on a Logic of "Better"', *Act. Phil. Fenn.* xvi (1963), p. 285 ff.

The instantial comparability principle affords obvious illustrations of this last point, and so does the consequence principle, so far as hypotheses of the appropriate form are concerned. For example, the wrongness of killing is greater than the wrongness of telling lies, where these are the only morally relevant features of what is done, if and only if the wrongness of your killing me is greater than the wrongness of your lying to me, where the same applies. Similarly the wrongness of your murdering ten people is at least as great as the wrongness of your murdering one or more people.[1]

Linguistic grammar is another important area of discourse in which support-, simplification-, and evaluation-functions all have a role to play. Let us suppose that an elementary grammatical generalisation, in the study of a particular natural language, has the form

(12) For any morphophonemic string x, if x has the structure R, x is a grammatical sentence.

Then for a hypothesis like (12), about English, in which the structure R consisted just in the occurrence of a pronoun, verb and noun, in that order, relevant variables must include such circumstances as pronominal case and verb-number, since, even though 'He eats fish' is grammatical, 'Him eats fish' and 'He eat fish' are not. The example is, of course, very much oversimplified. But to adopt any particular programme here for writing the grammar of natural languages would only serve to obscure the issue at stake. What is being suggested is a way to achieve the logically idealised reconstruction of inductive reasoning about grammatical statements, not the journalistic description of currently accepted grammatical theories. No doubt (12) represents a rather elementary type of grammatical hypothesis. Any systematic grammar would aim to state a relatively small number of postulates or rules from which a very large, and perhaps infinite, number of elementary hypotheses like (12) could be derived or generated. But in this respect the relation of a hypothesis like (12) to a systematic grammar presents an exact analogy[2] with the

[1] Those who suspect that paradoxes can arise here should bear in mind not only what has been said in §11 about criteria of testability and the derivation of all inductive support from canonical test-results, but also the fact that the inductive consequence principle applies to consequence-relations between whole conditionals like (11), not to consequence relations between their antecedents.

[2] Even if the structural description of a morphophonemic string is to be identified with the ordered set of rule-applications that generates it from a lexicon, we can still distinguish the perhaps infinite list of such descriptions from the rather small finite list of generative rules in the language. But the precise sense in which the latter list constitutes an axiomatisation of the language's grammar requires scrutiny: cf. L. Jonathan Cohen in *Psycholinguistic Papers*, ed. J. Lyons and R. Wales (1966), p. 165 ff.

relation of experimentalists' elementary hypotheses to a natural-scientific theory (cf. §9).

If U is a grammatical generalisation like (12), we can read the assignment of an ordinal value to the monadic evaluation function for U, viz. $e[U] = i/n$, as 'For any morphophonemic string x, if a full grammatical description of x is that x has the structure R, then x has just ith grade grammaticalness', where this has precisely the same sense as 'There is just ith grade support for the best supported hypothesis that implies, for any x, the grammaticalness of x if x has the structure R'. For example, suppose test t_2 on hypotheses materially similar to (12) varies pronominal case alone, while t_3 varies both pronominal case and verb-number, and so on. Then, the strings 'Him eat fish' and 'Him eats fish' would have just 1st grade grammaticalness, since the best supported hypothesis to uphold them would be the pronoun-verb-noun one and this would have just 1st grade support. But 'He eat fish' would have 2nd grade grammaticalness, and 'He eats fish' would presumably have nth grade grammaticalness, if there are n relevant variables.

Such a method of grading grammaticalness has three particularly important features. First, in virtue of its inductive syntax grade of support for grammatical hypotheses is both demonstrably incommensurable with any of the logical probabilities involved (§3), and also separately assessable from any of the statistical probabilities (§13). It follows that grade of grammaticalness, in the sense defined above, is likewise non-probabilistic. So this method of grading grammaticalness does not run foul of N. Chomsky's arguments against any kind of probabilistic model for syntactic structure.[1] Secondly, by making grade of grammaticalness depend on the grade of support that exists for a hypothesis about grammatical syntax we avoid the difficulties that arise when people try to make it depend instead on some measure of intelligibility.[2] In particular, syntactical sources of unintelligibility are not confounded with semantical ones,[3] and it is possible to treat the contribution of gram-

[1] *Syntactic Structures* (1957), p. 16 ff.

[2] Cf. L. Jonathan Cohen, 'On a Concept of Degree of Grammaticalness', *Logique et Analyse* 30 (1965), p. 141 ff., which criticises J. J. Katz's paper 'Semi-sentences', in *The Structure of Language*, ed. J. A. Fodor and J. J. Katz (1964), p. 400 ff.

[3] One important reason for retaining in linguistic theory a distinction between two types of restriction on sentence-formation—syntactical restrictions and semantical ones—is that some restrictions seem relatively stable and inflexible and others relatively fluid and malleable. The fluid, semantical restrictions often change more or less simultaneously throughout a culturally associated group of languages. But when the relatively stable, syntactic restrictions change at all, they tend to change in ways peculiar to individual languages. Secondly, intelligible communication normally requires a considerable level of redundancy in the message transmitted, and syntactic restrictions ensure the continued presence of some such redundancy even when the use of new meanings, or the communication of new facts, diminishes the amount of semantically derived redundancy.

maticalness to intelligibility as an issue that is open to psycholinguistic enquiry rather than as one that is closed by definition. Thirdly, we can compare the role performed by the ordered set of relevant variables in our analytical reconstruction, and that performed by the hierarchy of grammatical categories which Chomsky postulates.[1] If a procedure for projecting such a hierarchy from a grammar is a component in a language-learning device, then according to Chomsky, that device acquires the automatic ability to comprehend syntactically deviant strings when it learns the grammar of a language. Similarly, on our view, a knowledge of all the variants of relevant variables would enable any grammatically fluent English speaker to map the best supported hypothesis, H, vouching for a deviant string like 'Him eat fish', on to the simplest fully supported hypothesis (or hypotheses) that is a (or are) less simple version (or versions) of the same unmodified hypothesis of which H is also a version. Thus a language-speaker's ability to correct semi-grammatical sentences, which any linguistic or psycholinguistic theory must take into account, can now be seen as just one more manifestation of an intellectual ability that is also often manifested in other fields of inductive reasoning, wherever relevant evidence is available, viz. the ability to reduce a too simple version of a hypothesis (which in this case upholds the semi-grammatical sentence) to a more complex but better supported version. A child's, or native language learner's, tendency to produce such sentences is comparable with the tendency of an imprudent engineer, moralist, or judge, to act on conclusions from insufficiently qualified generalisations. In this, as in other respects (cf. §14, p. 126 f. above),[2] induction by variation of circumstance provides an explicit theoretical model for the inexplicit processes by which children learn the grammar of their native language.

Much is as yet unknown or controversial in linguistics, as in most other areas of scientific enquiry. But the proposal to use an inductive function for the appraisal of grammatical hypotheses, or the evaluation of grammaticalness, does not stand or fall with any particular theory

[1] 'Some Methodological Remarks on Generative Grammar', *Word* xvii (1961), p. 219 ff.

[2] Cf. also the way in which children tend to ascribe too wide a meaning at first to a noun they learn, and then gradually grasp the restrictions on its use. Children's tacit reliance, in their language-learning, on a principle of linguistic uniformity (e.g. every sentence-utterance instantiates some pattern of grammatical sentencehood, every actual word-use instantiates some pattern of meaningful word-use, and all like objects are named alike) may be compared with their tacit reliance, in judgements about their siblings, on a principle of fairness, impartiality or universalisability: the same underlying principle is at work. An inductive language-learning device is thus structurally isomorphic with an inductive ethics-learning device, as with an inductive device for natural-scientific discovery. Cf. L. Jonathan Cohen, 'Some Applications of Inductive Logic to the Theory of Language', forthcoming in *Amer. Phil. Quart.* vii (October, 1970).

about the nature of grammar. It is unaffected, for example, by doubt or controversy about the proper function of transformational rules, or about the nature and extent of linguistic universals. It makes no assumptions about whether a single support-function, based on a single hierarchy of grammatical categories, should—ideally—suffice for the study of all languages or for the study of each. It is even indifferent to the role of native-speakers' intuitions in the testing of grammatical hypotheses.

Indeed we can note here some interesting analogies between disputes about the nature of law and disputes about the nature of grammar.

Sociologists of law, and corporation lawyers who tell us how such-or-such a court is likely to decide certain cases, are like taxonomists of grammar who tell us which patterns of sentence construction are likely to be found. Realist philosophers of law, who tell us that the law is nothing but what the courts decide, are therefore like positivist philosophers of grammar, who tell us that all theorising about a language's syntax must fit the evidence provided by some corpus of actual utterances. Such determination to treat the subject-matter of legal or grammatical reasoning as a tissue of facts, rather than of rules, is quite compatible, as we have seen even in the case of natural science (§16), with the construction of unifying theories that appear as idealisations when contrasted with the actual phenomena. Saussure's distinction between language and speech need not be sacrificed by such a conception of grammatical theory.

But that distinction is reinforced if grammar is regarded instead as a tissue of rules, not of facts, and native-speakers' intuitions about grammaticalness are invoked as evidence. For then speech remains a merely *de facto* performance but language, at least in its grammatical aspect, is the rule-constituted competence of a correct native speaker. We can now compare the way in which native speakers are admitted to utter ungrammatical sentences sometimes, with the way in which one court may sometimes be declared by a higher one to have decided a case incorrectly. In both cases what is conceived to happen is the breach of a rule, rather than the interruption of a regularity. Also, just as innumerably many grammatical sentences are never uttered, so too innumerably many good causes of action never come to the attention of a law-court. Moreover, both legal and grammatical rules, thus conceived, are of a peculiarly hybrid kind, when contrasted with moral precepts or normative grammar. It is a matter of fact, not of normative right, that such-or-such a rule of generative grammar exists, i.e. forms part of the competence of certain native speakers (and not of others). But the rule determines how they ought to speak, not how they do speak. Similarly it is a matter of political fact, not of moral right, that such-or-such a legal rule exists, i.e. forms part of a certain legal system (though not, perhaps, of another). But the rule determines how cases ought to be decided, not how they are in fact decided. Again, a judge's action in awarding judgement is justified, not predicted, by his theoretical conclusion, and so he cannot come to confirm that conclusion. So too a native speaker may be in doubt whether or not to utter a certain sentence until his intuitions about other strings that fall under the same description assure him of its grammaticalness. But though he has now come to utter the sentence in question, it would obviously be circular for him to cite this performance of his own as further confirmation of his views about the grammatical rules that generate it. In short, generative grammarians are like analytical jurists, preferring to take the internal, rule-orientated point of view of the judge or native speaker, rather than the external, fact-orientated point of view of the corporation lawyer or linguistic taxonomist.

Thus, when we discover that inductive support- and simplification-functions are applicable to legal, moral and grammatical reasoning, as well as to experimental reasoning, we not only find further support for the view that certain fundamental principles are principles of logical syntax rather than peculiarities of reasoning about an empirical subject-matter. We also have some clarification of the similarities and differences that exist between these other areas. Moreover we can find yet further areas of thought to which some non-standard interpretation of the inductive functors would apply. For instance, G. Pólya, developing the hints of some older mathematicians such as Euler and Laplace, has argued that induction has not only a heuristic role to perform in mathematical reasoning, but is also a mode of justification that is useful where a demonstration is not available, and may be assessed by a credibility-function.[1] He himself regarded such a function as determining a logical probability, albeit a non-quantitative one. But some of his examples of the different kinds of tests to which a mathematical conjecture may be submitted exhibit the characteristic features of induction by variation of circumstance,[2] and for such a mode of reasoning his own credibility-function would need to be replaced by appropriate inductive support- and simplification-functions.[3]

[1] *Patterns of Plausible Inference* (1954), p. 109 ff. Proof by recursive induction, of course, counts here as a form of demonstration. I. Lakatos, 'Proofs and Refutations', *Brit. Jour. Phil. Sci.* xiv (1963–4), p. 1 ff., p. 120 ff., p. 221 ff., and p. 296 ff., has developed Pólya's ideas in a way that, amongst other things, exhibits the applicability of non-standard methods of inductive reasoning in mathematics—e.g. what Lakatos calls 'exception-barring methods'.

[2] E.g. the polyhedron example in *Induction and Analogy in Mathematics* (1954), p. 35 ff.

[3] I have not specified separately the criteria of testability for legal, moral or grammatical hypotheses, since these criteria must be the same as the three criteria for experimental

Beyond a certain point, however, less and less is gained by multiplying the evidence for claiming that certain structural features of inductive functions are principles of logical syntax which remain invariant from one subject-matter to another. It is more important to show that these principles satisfy yet further criteria of adequacy. Indeed, if criteria of adequacy for a philosophical analysis of explication are derived, as suggested in the Introduction (p. 1 f.), from the difficulties that have floored previous philosophical discussions of the same or similar subjects, we are entitled to treat such a criterion as a test in relation to a relevant logico-philosophical variable. So if an analysis of inductive syntax has, as it were, survived the manipulation of several variants of a relevant logico-philosophical variable—e.g. several semantical reinterpretations of the functors concerned—we shall do best to seek further support for it by taking yet another relevant variable into account.

hypotheses that were specified in §11. So far as mathematical hypotheses are concerned, however, we must obviously weaken the requirement (of §11) that neither a testable hypothesis itself, nor the antecedents or consequents of its substitution-instances, may be propositions that are logical consequences or contradictions of accepted mathematical postulates. What is required here is merely that they should not be logical consequences or contradictions of accepted mathematical postulates in the same field, i.e. of postulates from which a demonstration would, ideally, be available. The equivalence and consequence principles must be interpreted correspondingly.

VI

The Justifiability of Induction

§19. Does it make sense to ask for a justification of inductive inference?

It has been argued so far that certain patterns of logical syntax are deeply entrenched in actual discourse. But are such patterns genuinely patterns of support, or valid inference? Is what we have been calling 'inductive support' a justifiable ground for inference? Some philosophers have argued that it does not make sense to ask for a justification of induction, because it is analytic that a valid inductive inference is an inference of such-or-such a kind and because there are no standards by reference to which inductive standards could be justified. This argument is not refuted by the claim that it represents an analogue of what Moore called 'the naturalistic fallacy'. But it is refuted by the objection that even if H is analytic we can always ask whether our concepts should be such that it is so. Hence a useful sense imputable to requests for a justification of induction is the question: what is our title to use a term like 'valid' in relation to inductive inference? How can this ordinary usage be ratified by philosophical argument? What analogies are there between valid cases of deductive inference and reputedly valid cases of inductive inference, such that the use of the same word 'valid' in both cases is more than just a homonym? Hume's scepticism about inductive reasoning was based on an argument from disanalogy here: to answer him we must find analogies that he overlooked.

If the criteria of adequacy for a philosophical analysis of inductive syntax are conceived to represent the relevant variables for testing it, the present analysis has so far survived encounter with five such variables. First, it turned out to be as well instantiated—*mutatis mutandis*—by experimental support for causal hypotheses, correlational generalisations and scientific theories, as by experimental support for more elementary hypotheses (§§9–10). Secondly, it was not hit by paradoxes like those that have been proposed by Hempel and Goodman (§11). Thirdly, it was as well instantiated by experimental support for statistical generalisations as by that for non-statistical ones (§§12–14).

Fourthly, it was as well instantiated by simplification-functions as by support-functions (§16). Fifthly, the analysis was as well instantiated by legal, moral and grammatical, as by experimental, reasoning (§§17–18).

No doubt many other types of appraisal are in common use besides those conforming to the logical syntax of an inductive function. Simplicity, for example, can no doubt be assessed in many different ways, and so can grammaticalness. Evaluations of moral wrongness can be conducted by the counting or measurement of utilities, as well as by the method described in the previous section (§18). Even range-theoretical functions may have some genuine application, e.g. to predictions about the winner of a chess-tournament.[1] What is claimed therefore for the present analysis is not at all that it articulates the only possible syntax for support, simplification or evaluation. Instead it articulates one syntax for them which is deeply entrenched in actual patterns of discourse and which on historical grounds, traceable originally to the genius of Francis Bacon (§14), has a prerogative title to the appellation 'inductive'. But though previous chapters have shown the entrenchment of this syntax in commonly used patterns of discourse the question may still be asked: are such patterns rightly entrenched? Are they mutually consistent, and would E really support H even if it were true, that, according to the appropriate support-function, $s[H, E] = n/n$? The problem of demonstrating consistency between our various syntactic principles requires a formal treatment, and will be dealt with in §22 below. The present chapter will concentrate on the more philosophical issue of whether the evidence of inductive test-results can really support a universal hypothesis. I shall treat this problem, however, as posing a criterion of adequacy that any analysis of inductive reasoning should satisfy, rather than as providing an opportunity for an academic exercise in the proof or disproof of scepticism.

It has sometimes been said or suggested, not only that all actually attempted justifications of induction have failed, but also that no such attempts can possibly succeed because no non-trivial sense can be given to the interrogative sentence 'How is induction justified?' Here, however, it will be argued that a philosophically fruitful sense can be given to this purported question and that up to a certain point, at least, it is answerable. Not that every sceptical doubt about induction will thereby be laid at rest, since the type of justification to be proposed is quite different from the types that sceptics have previously been supposed to require. To rebut the sceptics philosophers have either tried to *validate* inductive argument, as a logical form, by claiming that though it invokes special

[1] R. Carnap, *Logical Foundations of Probability*, p. 382 f.

principles, such as the uniformity of nature, these are all independently certifiable. Or they have tried to *vindicate* the policy of inductive enquiry, as a heuristic method, by claiming it a suitable means to desired ends, e.g. because, allegedly, it is self-correcting. But it turns out a mistake to suppose, as has sometimes been supposed,[1] that validation and vindication, so conceived, are the only forms of justification worth considering for induction. The problem of induction can be given a new lease of life—though not by offering new answers to the old questions, which may well be answerable, but by asking a new question.

Two main arguments[2] have been offered for the thesis that no non-trivial sense can be given to purported requests for the justification of induction. The first argument—let us call it the argument from ordinary language—is that to ask for a proof that it is reasonable to place reliance on inductive procedures is quite pointless, since proportioning the degree of one's conviction to the strength of the inductive evidence (construed as the number and variety of observed cases) is just what 'being reasonable' ordinarily *means* in such a context. The second argument—let us call it the argument from the need for standards—is that it is senseless to ask whether the use of inductive standards is justified unless we can say to what other standards we are appealing for this justification, just as it makes no sense to inquire in general whether the law of the land as a whole is or is not legal since there are no other legal standards to which we can appeal.

On the other hand, the thesis that it is senseless to ask for a justification of induction has been criticised[3] on the ground that the argument from paradigm cases in ordinary language does not have the same efficacy for questions about evaluative terms like 'good' or 'valid', as for questions about non-evaluative ones like 'solid'. We cannot query that 'solid' means something like 'of the consistency of such things as desks'. But we are always in a positition to ask, it has been said, whether there are any

[1] J. J. Katz, *The Problem of Induction and its Solution* (1962), p. 24 ff. Cf. V. Kraft, 'The Problem of Induction', in *Mind, Matter and Method* (Essays in Honour of H. Feigl), ed. P. K. Feyerabend and G. Maxwell (1966), p. 306 ff. It was H. Feigl who, in 'De Principiis Non Disputandum ...?', *Philosophical Analysis*, ed. M. Black (1950), p. 119 ff., first used the terms 'validation' and 'vindication' to classify all justifications, including justifications of induction, into two general kinds, depending on whether the justification is of a claim to knowledge or of an action, respectively.

[2] Cf. P. F. Strawson, *Introduction to Logical Theory* (1952), p. 256 f.; P. Edwards, 'Bertrand Russell's Doubts about Induction', in *Logic and Language*, ed. A. Flew (1951), p. 68 ff.; and A. J. Ayer *British Empirical Philosophers* (1952), p. 26 f. I am not concerned here with the exact formulations these philosophers adopted, but with two interesting and important arguments which, even if they have never been advanced by Strawson, Edwards or Ayer in just that form, are certainly worth considering in their own right.

[3] J. O. Urmson, 'Some Questions Concerning Validity', in *Essays in Conceptual Analysis*, ed. A Flew (1956), p. 120 ff.

good reasons to accept certain standards for grading apples, and similarly we are always in a position to ask whether there are any good reasons to accept the principles by which we count as valid those arguments, inductive or ethical, which we do count as valid. 'Valid' is an evaluative term and so must always have some other element in its meaning than the mere description of a certain structure of reasoning. Hence it cannot be just analytic, it has been argued, to say that an argument is valid in virtue of its having such a structure. It may be folly to doubt the principles of induction, but it certainly is not senseless.

But this criticism is much weaker than it seems and by no means settles the issue. First, it gives no answer at all to the argument from the need for standards, since it does not say how we can tell whether something is a good or bad reason for accepting the principles by which we count as valid those inductive arguments that we do count as valid. Perhaps you can judge my ethical standards by appealing to your own. But how do you judge the inductive standards that seem common to all reasonable men? This question is crucial for any attempt to invoke here an analogue of Moore's argument against proposals to define 'good' in terms of some describable characteristic.[1] Moore argued, with at least some plausibility, that we always understand very well what is meant by doubting whether the chosen, defining characteristic is itself good, and hence that no such proposed definition for 'good' can succeed. But the trouble with any attempt to argue analogously in relation to inductive validity is that it is patently question-begging. The question at issue is just this: do we really understand what is meant by doubting whether reasoning is valid in virtue of conformity to ordinary inductive standards, if, as is argued, we cannot cite other standards by which to judge the value of ordinary inductive ones?

Secondly, and perhaps less importantly, the argument from ordinary language may easily be restated in a modified form so that it is not open to criticism on the ground of failing to distinguish adequately between the evaluative and descriptive elements in the concept of validity. The premiss might be advanced, for instance, that it is analytic that an inductively valid inference is an inference to a generalisation from premisses that report an appropriate number and variety of cases in which the generalisation has held and none in which it has failed to hold, and that it is also analytic that an inductively invalid inference is an inference to a generalisation from premisses that report an unfavourable case, or at least no favourable cases, or very few of these, or mostly of the same kind. Such a premiss would rest on the commonly accepted assumption that an evaluative term like 'valid', 'invalid', 'good', or

[1] G. E. Moore, *Principia Ethica* (1903) p. 15 f.

'bad', may be supposed to have a descriptive implication, in certain contexts, that derives from its criteria of application, especially where no alternative criteria are envisaged for those contexts. This premiss would not imply that the converse propositions are also analytic, i.e. that it is analytic that inferences of these two kinds are valid and invalid, respectively. But the premiss would clearly imply (in its first half) the pointlessness of asking for a proof that an inductively valid inference is an inference to a generalisation from premisses that report an appropriate number and variety of favourable cases and no unfavourable ones; and it would also imply (by contraposition of its second half) the pointlessness of asking for a proof that an inference of this kind is not inductively invalid.

Nevertheless, even if the propositions in question are analytic, it is still possible to discuss whether the phrase 'is an inductively valid inference from' (or connected locutions like 'is an inductively reasonable inference from', 'gives full inductive support to', etc.) ought to be used in a way that gives rise to their analyticity. The fact is that in addition to discussing how terms are actually used we can also discuss the merits of those usages. Sometimes this discussion is purely stylistic and has reference only to modes of expression within one particular natural language. Sometimes, with more philosophical interest, the discussion is concerned with the merits of linguistic usages that are invariant under translation from one language to another. The traditional philosophical problem of universals, for example, may be fruitfully interpreted as the question whether general words (numerals, species-names, property-terms, etc.) should be used, in this or that area of human thought, on the assumption that they denote abstract entities, or on the assumption that they denote mental constructs, or on the assumption that they are merely oral or notational tools.[1] Similarly it might be asked whether the term 'number' is justifiably used in a sense that makes it true to call zero,[2] or the square root of minus one, a number, or whether the term 'geometry' is justifiably used in a sense that will make it true to call Lobachevsky's postulates a geometry. But what guarantees this philosophical type of discussion to make sense is not at all that it is concerned with evaluative rather than non-evaluative terms. It is not the *kind* of term, meaning or concept that matters here, but the way in which we conceive of our concepts. So long as we think the same concept capable of more than one form or role[3]—for example, lacking certain necessary conditions in one form or role that it has in another—we are always in a

[1] Cf. L. Jonathan Cohen, *The Diversity of Meaning*, 2nd ed. (1966), p. 131 ff.

[2] The ancient Greeks knew of no such number, and its introduction into European mathematics in the tenth century was highly controversial.

[3] The thesis that concepts are not capable of this is criticised by L. Jonathan Cohen, op. cit., p. 103 ff.

position to discuss whether one form or role is justifiable by reference to another, and therein lies the proper answer to the argument from ordinary language. Moreover one good way of doing this is to point to the nature and extent of the analogies that exist between what falls under the concept in one form or role and what falls under it in the other, and therein lies an answer, as will be shown in detail below, to the argument from the need for standards.

Thus the premiss of the argument from ordinary language (in its modified form) can readily be granted. Very much the same thesis would have been accepted by the arch-sceptic himself, David Hume. Hume frequently applies the word 'reasoning' to the mental processes by which we come to form beliefs about particular kinds of cause and their effects and to expect future events of certain kinds to resemble past events of those kinds,[1] and he listed rules that he thought 'proper to employ' in such reasoning.[2] Only on some occasions does he prefer to make his point instead by saying that constant conjunction can never be an object of reasoning or that we have no reason to draw any inference concerning any object beyond those of which we have had experience.[3] So Hume would presumably not have wished to deny that in a common or vulgar sense of the word induction may be termed a process of 'reasoning' which, when it works properly, is an inference to a generalisation from premisses that report an appropriate number and variety of favourable cases and no unfavourable ones. What he is maintaining is rather that this is a radically different process from demonstrative reasoning. His sceptical thesis about induction[4] boils down to a thoroughgoing insistence on the pervasive lack of analogy between, on the one hand, reasoning from premisses to logically implied conclusions, and, on the other, reasoning—vulgarly so-called—from premisses about the already observed, to conclusions that embrace the as yet unobserved:—the former is a product of thought, the latter of custom; the former justifies certainty, the latter not; the former cannot be rejected without self-contradiction, the latter can; and so on. Hence Hume may be taken to be arguing in effect that, insofar as valid deduction is taken as the paradigm of what is referred to by the term 'valid reasoning', the use of that term to refer to reputedly valid examples of induction is an extended and ill-justified one, since there are few analogies between the two processes of thought.

It will be argued here, however, that there are important structural

[1] E.g. *Treatise*, ed. Selby-Bigge (1888), pp. 172, 173, 175, 177, 179, 181, 183.
[2] Ibid., p. 173 ff.
[3] Ibid., pp. 170 and 139, respectively.
[4] As distinct from what he says later about total scepticism.

analogies, overlooked by Hume, which do afford some justification for this extended use of the term 'valid reasoning'. So, since arguments of such a kind can intelligibly be opposed to Hume's, it follows that at least some clear sense can be given to the question whether or not induction is justified. Both arguments that impugned the meaningfulness of that question turn out to be inconclusive. Not only does the argument from ordinary language overlook questions about the rationale, as distinct from the facts, of ordinary usage. But also the argument from the need for standards tends to overlook the standards that are commonly considered relevant to such issues, viz. the nature and extent of the analogy between one or more cases that fall indisputably under the term in question and one or more cases that would fall under it in the allegedly extended sense. Thus it makes excellent sense to ask whether laws of behaviour like those prevailing in a certain territory are or are not legal, if this is intended as an enquiry whether certain rules of social organisation are such that the term 'legal system' can justifiably be used in a sense that would embrace them in its reference. We may have a detailed account of what such laws enjoin and how they operate and yet still be in doubt whether they are justifiably termed 'legal,' or only 'moral'—just as we may well feel that we know exactly how induction proceeds, and yet still be in doubt whether it is justifiably termed reasoning or not. The doubt is not about the laws themselves, but about the concept of a legal system; and the doubt may be resolved by examining the nature and extent of the analogy between such a pattern of social organisation and some well-established legal system. Philosophers have long since learned how to handle questions of this kind—is the square root of minus one a number? are there non-Euclidean geometries? etc.—and by assimilation to such issues the problem of induction is rendered into a familiar and manageable form.

It may perhaps be objected that it is not really analytic that valid deductive, or inductive, inferences are of such-or-such a kind. I.e., it may be urged that we are not really involved here with two differently analysable senses of 'valid reasoning'—whether a restricted and an extended one, or two non-overlapping though co-ordinate ones—but rather with two different sets of criteria, one deductive and the other inductive, for applying a single concept of valid reasoning. Something like this is often said in relation to other evaluative terms, like 'good', for example, where there are said to be many different sets of ethical and non-ethical criteria for applying one and the same term. But the substantive point at issue is unaffected by any such re-characterisation. The analytic-synthetic issue here, as often elsewhere,[1] is of little importance.

[1] Cf. §6, p. 46, and the footnote on that page.

We either ask how one meaning is legitimated by another since, as Aristotle[1] put it in the case of different kinds of good, it does not look as though we are dealing with merely accidental homonyms. Or we ask how the use of one set of criteria for the application of a particular evaluative term legitimates the use of another set for the application of the same term. In either case we shall do well to seek analogies between what is referred to, as Aristotle suggested in the case of goodness. But we shall not be aiming to reduce inductive meaning or criteria to non-inductive ones by showing that the former is or are just a special case of the latter.

Indeed, the older conceptions of the problem of induction seem to regard some such reduction as being required, and the older methods of trying to justify induction may be viewed as trying to achieve this. Validations normally sought to present inductive reasoning as relying solely on the logical or mathematical criteria of valid reasoning that are invoked in deduction from non-controversial premisses. Vindications sought in effect to rely on a third set of criteria for reasonableness, viz. criteria for the reasonableness of performing certain actions as distinct from the reasonableness of making certain inferences—practical reasonableness as distinct from deductive reasonableness. In contrast with those older, reductive methods, the present project begins from the assumption that there are distinctive, non-reducible criteria of inductive reasonableness and that therefore the older, reductive methods are misplaced. What are now to be sought are arguments that will legitimate or ratify—ratify with the seal of philosophical approval—the ordinary vulgar convention by which the terms 'is validly reasoned from', 'is a reasonable inference from', etc., or at least the terms 'gives full support to', 'gives some support to', etc., are taken to be assignable in accordance with these distinctive criteria.

It is convenient to call this objective the 'ratification' of induction, so as to distinguish it from validation or vindication. But ratification is not directed, as validation and vindication often are, towards justifying an increased level of certainty in the conclusions of valid inductive inferences. Hume's problem is not a problem about people who are deficient in such certainty and require philosophical reassurance. Such people are Cartesian sceptics, not Humeian ones. Hume denied, in effect, that any people of this kind exist, since inductively based expectations were natural, on his view, to both human beings and animals. Hence it is no answer to Hume to say, as he would undoubtedly have admitted, that increased confidence in an inductive conclusion is to be gained by obtaining more and better evidence, not by philosophical argumentation. Hume's scepticism is not about whether the sun will rise tomorrow, but

[1] *Nicom. Eth.*, I, vi.

about whether validity can be ascribed to any inference from past observations that it will rise to-morrow. The ratification of induction is directed against scepticism of the latter kind, not the former. It has the typically philosophical purpose of elucidating or explicating why it is that terms like 'valid', 'reasonable', 'support', etc. are applicable to inductive procedures.

Arguments from analogy are far from being the only ones capable of ratifying ordinary usage.[1] But they are peculiarly appropriate to the present problem, because they counter Hume's appeal to *dis*similarities between deduction and induction. Nor must all analogical ratifications of reasonableness employ the same paradigm. It is conceivable, for instance, that to answer some sceptics one should seek to ratify deductive reasonableness by reference to analogies with practical or inductive reasonableness, rather than to ratify either of the latter by reference to analogies with the former. But in relation to scepticism about inductive reasonableness deducibility is given as a paradigm by the terms in which Hume originated the problem. Also, I shall assume here that the serious philosophical problem at issue is how to ratify induction from variety of circumstances, since enumerative induction, i.e., induction from multiplicity of instances, is to be regarded as a form of statistical estimation that is vouched for (when duly qualified in relation to confidence interval) by the mathematical facts underlying Bernoulli's theorem (cf. §12, p. 109 ff. above).[2]

Perhaps someone will object to the whole enterprise, on the ground that it must be circular to seek an analogical justification for induction, since analogical argument is itself a form of inductive reasoning (cf. §17, p. 164 f.). An analogical justification for induction, it will be said, must first generalise inductively from the paradigm evidence provided by deducibility and then show that valid induction itself falls under this generalisation.

But there is no real circularity here. It would certainly be viciously circular to propose an analogical justification of philosophical reasoning by analogy. What will be attempted here, however, is a philosophical justification, by analogy, for the experimenter's form of inductive

[1] See L. Jonathan Cohen, *The Diversity of Meaning* 2nd ed. (1966) p. 118 ff. for some others. It should also be admitted that against a Cartesian sceptic, as distinct from a Humeian one, the only procedure to employ may well be the one that is mentioned by Strawson and Edwards, viz. the presentation of more and more evidence. This should work eventually, if Hume was right about human nature.

[2] Cf. also D. Williams, *The Ground of Induction* (1947). But Williams made the mistake of supposing that when he had elucidated the nature of so-called enumerative induction he had solved the whole problem. He almost entirely ignored the importance of using systematic tests to allow for the effects of variety and heterogeneity in the domain of a generalisation; and so he missed the central problem—how to justify reliance on such tests.

reasoning. Hume raised his problem about the standard, experimental form of inductive reasoning, not about the other forms (legal, ethical, grammatical, mathematical, philosophical, etc.) which happen to be discussed in the present book. Now, that problem has a peculiar acuteness because we cannot take the validity of experimental reasoning to be just a product of human decisions, though the validity of inductive reasoning in at least some of its other forms might be taken to be sanctioned in one way or another by human decisions. The adoption of common law, for instance, with its respect for judicial precedents and the doctrine of *similia similibus*, is a matter for human choice; so that the success of legal induction, if properly executed, is guaranteed, as it were, by that choice, whereas we do not choose, or create, the uniformities studied in experimental science. In the same way the use of just one language at a given time, with the corollary that like things should be named alike, is also a matter for human decision; and such a decision sanctions analogical arguments about correct nomenclature, as in the ratification of experimental induction. There is thus no risk of circularity here. Indeed the ratification proposed for experimental reasoning is adaptable, *mutatis mutandis*, for any form of non-philosophical reasoning by induction. Moreover our title to use analogical reasoning in support of this proposal is underwritten by Hume's use of it, in a negative sense, to support his scepticism. For Hume's arguments from disanalogy to have force they require the assumption that things should be named or conceived alike if and only if they are alike.

§20. The analogical ratification of induction

If analogies between induction and deduction are to be found by comparing the all-or-nothing relation of deducibility with some varying relation of inducibility, then there are necessary-but-not-sufficient conditions of the former that cannot belong to the latter. Carnap's theory would make all the necessary-but-not-sufficient conditions of the latter belong to the former, but his theory is otherwise unsuitable as a basis for ratifying induction. In any case the comparison should rather be between deductive inferences to conclusions about what is logically true, on the one side, and, on the other, inferences warranted by accepted inductively supported conditionals to conclusions about the existence of such-or-such a grade of inductive support for a certain hypothesis. All the necessary-but-not-sufficient conditions—i.e. the logical syntax—of the latter turn out to belong also to the former. It remains to deal with the objection that these syntactic analogies do not favour any one set of inductive criteria, or ordered set of experimental tests, as against any other.

> But it is inevitable that the proper test to use in a particular field should itself be a matter for inductive argument, and this should not be regarded as a bar to the ratification of induction.

The ratification of induction requires the discovery of analogies between deduction and induction that can justify us in including the latter within the domain of certain terms, such as 'valid reasoning', which are commonly applied to the former. But what kind of properties should these analogies collate? Presumably such properties should be logical, rather than psychological, if questions of validity and justification are at stake. But there are obviously going to be some properties that belong exclusively to deductive inference, such as logical inconsistency between the truth of the premisses and the falsehood of the conclusion. We shall no more want to regard induction's lack of these properties as an argument for not holding inductive inference reasonable, than we shall want to take the absence of a law guaranteeing the Hanoverian succession as an argument for not holding the Napoleonic code to be a legal system. How then should we characterise the properties we are interested in and the properties we are not?

Let us call a 'logical property' any property that is definable within the vocabulary of formal logic alone: e.g. transitivity, symmetry, etc.[1] Presumably it is some of these properties that are at stake if logical analogies are to be found. But which? We certainly need to discover analogies that hold in every case, not just in some. Non-necessary features of induction or deduction do not concern us. So let us call 'a necessary logical condition' of an n-adic property or relation, where $n \geqslant 1$, (or 'necessary condition' of it for short) any logical property of it that is definable without the use of existential quantification (or of the denial of universal quantification) over non-logical variables. E.g. transitivity and invariance under contraposition are necessary conditions of deducibility according to standard, classical logic, but neither symmetry nor asymmetry nor non-symmetry are. But some necessary conditions of deducibility will also be sufficient conditions and these again do not concern us. Let us call a 'sufficient logical condition' of an n-adic property or relation R, where $n \geqslant 1$, (or 'sufficient condition' of R, for short) any logical property of R such that if R′ has this property then R′ also belongs to (or relates) all and only the things which R belongs to (or relates). E.g. it might be held that the property of relating any two statements of which the second contradicts denial of the first is a sufficient condition of deducibility. So the position may be summed up by saying that in searching for ratificatory analogies between induction and

[1] Cf. B. Russell, *The Principles of Mathematics*, 2nd ed. (1937), p. 8.

deduction we are primarily interested in necessary-but-not-sufficient conditions.

But which way can the analogy run? Do all the necessary-but-not-sufficient conditions for deduction hold also for induction, or vice versa, or is the analogy incomplete in both directions? It will be convenient to show first that the analogy is incomplete in the former direction. But before doing so, we must clarify a little further the terms of the comparison.

A problem is created by the fact that inductive support is a matter of degree, not of all or nothing, whereas in deductive logic there are no gradations. So what precisely do we compare, property by property? Superficially the most obvious comparison to make is between the all-or-nothing deducibility of one proposition, H, from another, E, and the varying inducibility, as it were, of H from E. On this view inductive support is fundamentally—i.e. when articulated in logically primitive terms—a type of dyadic relation between propositions, to be compared with logical implication. For example, Carnap has traced (though not for ratificatory purposes) some of the analogies between deductive logic as dealing with the relation of total inclusion between ranges—the range of a sentence *s* in a language-system L is the class of those state-descriptions in L in which *s* holds true—and inductive logic as dealing, allegedly, with the relation of partial inclusion.[1] But, whether or not we adopt a range-theoretical analysis of inductive support, this basis of comparison would not admit of a complete analogy from deduction to induction. For there are necessary-but-not-sufficient conditions of classical deducibility, such as transitivity, or invariance under contraposition, that are not also necessary conditions of inductive support.

It is perhaps easiest to see this in the case of transitivity. If E gives a certain grade of inductive support to H, and H gives the same grade of support to H′, we cannot infer that E also gives this grade of support to H′: we can infer only that E gives this grade of support for there being evidence of this grade of support for H′. But contraposition fares just as badly. If inducibility is invariant under contraposition, then, where E reports a favourable test-result for a universal hypothesis U, the negation of U will give just as much support to the negation of E—and thus to the proposition that the test-result was in fact unfavourable—as E itself gives to U. And this seems very paradoxical since we feel that the falsehood of U, if it is false at all, may well be due to the effect on it of other variables than those so far manipulated in our tests. If a previously successful hypothesis turns out, under a more thorough test, to be false, is there really now just as much support for saying that there must have

[1] *Logical Foundations of Probability*, p. 199 ff.

been some error in the reports of previous tests? After all, scientists are sometimes just unlucky in the level of thoroughness they impose on their tests, and not at all inaccurate or dishonest in their reports of those tests. If therefore we refuse to accept this kind of retrospective discrediting of perfectly respectable work, it looks as though we must reject inductive contraposition.

No doubt some putatively sufficient conditions of deducibility are controversial in respect of certain of their consequences, such as the deducibility of any statement from an inconsistent one. But about the necessary-but-not-sufficient conditions of deducibility there is relatively less dispute, and it may seem a little disappointing that apparently not even all these relatively uncontroversial, necessary-but-not-sufficient conditions of deducibility, such as transitivity and invariance under contraposition, have plausible analogues for inductive inference. No doubt some of the disanalogies might be avoided by adopting some non-standard, non-classical conception of deducibility. For example, acceptance of an intuitionist formal logic, like Heyting's, might remove the ground for disappointment that inductive inference is not invariant under the converse of contraposition: i.e. under the move from $-A \rightarrow -B$ to $B \rightarrow A$. But not all the disanalogies can be plausibly removed in this way without, as it were, putting the cart before the horse. If we modify our conception of deducibility each time we encounter a disanalogy with inducibility we shall end by destroying the whole basis for ratification. Our concept of deducibility will lose the paradigmatic value that it acquires from well-nigh universal acceptance. So it seems pointless to try and accommodate thus the disanalogies in respect of straightforward contraposition and transitivity.

Now the incompleteness of analogy, on this basis of comparison, is a serious bar to ratification. Moveover, not only do some important logical properties belong to deducibility but not to inducibility: there is also the trouble that some of the necessary-but-not-sufficient conditions of deducibility are necessary conditions of relations that can exist between two propositions which are such that no kind of reasonable inference is thought possible from one to the other. For example, transitivity and invariance under contraposition are also properties of material (i.e. truth functional) implication, even though not of inductive inference, and we do not think it possible to infer, in any way at all, from the true proposition that Julius Caesar is dead to the true proposition that the earth is not flat. So, if we do not take the possession of some, but not all, of the necessary-but-not-sufficient conditions of standard, classical deducibility to warrant extending the term 'support', or 'valid reasoning', to cover implications or inferences in such cases, it may well be doubted whether any analogy that does exist in the case of inductive inference supplies any ratification for the ordinary use of such a term in that case. Conceivably someone might want to argue that the necessary-but-not-sufficient conditions of deducibility which are also necessary conditions of the reasonableness of an inductive inference are more important than those that are also necessary conditions of, say, truth-functional implication, and thence he might try to argue that the former analogy carries greater weight than the latter in justifying the application of terms like 'valid reasoning'. But it is difficult to see how such an argument could be based on any better foundation than a *petitio principii.*

So perhaps we should look at the problem for a while in the other direction. Instead of seeking analogues in induction for the necessary conditions of deducibility, perhaps we should seek to devise a systematisation of the necessary-but-not-sufficient conditions of inducibility that would exhibit those of deducibility (or at least of the deducibility of certain types of conclusions) as a special case. Part of the logical syntax of logical implication (or at least of certain types of logical implication) would thus be included as a sub-system within the logical syntax of inductive support, and we would then be sure that conditions necessary-but-not-sufficient for induction are also necessary-but-not-sufficient for deduction (in such cases). Instead of reducing induction to deduction, as is done in attempts at validation, we would rather be reducing deduction (or at least the deduction of certain types of conclusions) to a form of induction. The sense of 'valid reasoning' would appear as if extended to cover induction by a process analogous to that by which, for example, the sense of 'number' has been extended to cover fractions, surds, etc. so that the positive integers, to which the term was once confined, now figure as just a special case.

A system of this nature was certainly proposed both by Carnap's theory of symmetrical c-functions and by his later system of General and Special Axioms.[1] Even though Carnap's theories were offered just as explications of induction, they nevertheless provide a *prima facie* basis for its ratification, if ratification proceeds in the direction that we are now considering—since logical implication emerges as the relation of total inclusion between ranges. However, there are good reasons why Carnap's theories cannot in fact provide a basis for the ratification of induction.

First, as has already been argued, inductive support is not measured by a logical probability, whereas Carnapian c-functions assign logical probabilities. Perhaps it will be objected that even if inductive inference, in the sense of inference from experimentally established premisses,

[1] *Logical Foundations of Probability*, p. 483 ff.; and 'Replies and Systematic Expositions', in *The Philosophy of Rudolph Carnap*, ed. P. A. Schilpp (1963), p. 973 ff.

cannot be ratified along Carnapian lines, there may nevertheless be some other mode of inference that can be so ratified. But then a second point comes into play. The trouble is that there are infinitely many symmetrical c-functions, assigning different degrees of confirmation in relation to any given inductive inference. For example, $c^+[Rc, Ra \& Rb] = c^+[Rc, Ra]$, while—more satisfactorily—$c^*[Rc, Ra \& Rb] > c^*[Rc, Ra]$. But the ratificatory analogies with deducibility hold equally for c^*, c^+ and all other symmetrical c-functions. On their basis it seems just as proper to extend the meaning of 'valid reasoning' to cover inferences based on c^+ as to cover those based on c^*. Correspondingly, in the system of General and Special Axioms, a suitable value has to be chosen for the parameter λ.[1] If our ratification is as indiscriminate, or incomplete, as this it can hardly be worth much.

However there is a good reason why our quest for a ratification of induction has so far been unsuccessful. We have been assuming that the correct comparison to make, in our search for analogies, is between the all-or-nothing deducibility, and the varying inducibility, of one proposition from another. We have been assuming, in effect, that inductive support is fundamentally a type of dyadic relation or dyadic operator. But that assumption blocks normally accepted patterns of detachment from support-assessments. It seems to state that the relation between E and H, when $0 < s[H, E] < n/n$, is a connection of some kind that is weaker even than that of material (truth-functional) implication, since we certainly cannot infer the truth of H from the truth of E plus the fact that E gives H some low-grade support. Hence special rules need to be given about the circumstances in which this kind of connection does entitle us to detach a conclusion. Carnap, for example, allows us—given E and $c[H, E] = i$—to detach a conclusion that i-degree confirmation exists for H if and only if E states all the evidence available.[2] But such a requirement of total evidence has no analogy in the case of deductive logic, as Carnap himself points out,[3] and if it were indeed indispensable for inductive logic it would stand as a major obstacle to ratification. The logical syntax of inductive support could contain no analogue for the *modus ponens* of deductive logic, because the alleged requirement of total evidence in inductive logic would have no analogue in relation to deduction.[4]

[1] On the strength of this ratification it would be plausible to claim that these axioms are, or should be regarded as being, analytic. But the value chosen for λ must then be regarded as a synthetic *a priori* truth, if true at all.

[2] *Logical Foundations of Probability*, p. 211.

[3] Ibid.

[4] For the *acceptability* of a conclusion, as distinct from its derivability from given premisses, the situation is quite different: see §1, p. 8 f.

In ordinary reasoning, however, as was argued in §8 above, no special rules are supposed to cover the detachment of $s[H] \geqslant i/n$ from E and $s[H, E] \geqslant i/n$. Experimenters are not thought rash just because they draw inferences from their own test-results. Test-results are assumed to be repeatable, so no requirement of total evidence operates—no insistence on a record of all actual performances of the test. It follows that $s[H, E] \geqslant i/n$ must be regarded as equivalent to the minimum suitable premiss from which, when conjoined with E, we can detach $s[H] \geqslant i/n$. Rewriting $s[H] \geqslant i/n$ as $\Box^i H$ we can state this result in the form: $s[H, E] \geqslant i/n$ is equivalent to a conditional proposition

(1) If E is true, so is $\Box^i$H.

But what kind of conditional proposition must (1) be? Obviously, in the light of §§5–8, (1) is not logically true, and to regard it as being truth-functionally true would lead to intolerably paradoxical values for dyadic support-functions whenever E was false. In fact the truth of propositions like (1), or its equivalent '$s[H, E] \geqslant i/n$', is normally accepted on the strength of a competent experimenter's inductively fully-supported beliefs about what variables are relevant to the particular class of materially similar hypotheses over which the support-function s is defined and in what order they are relevant. So we can replace (1) by

(2) s[If E is true, so is $\Box^i$H] $\geqslant n/n$,

which may in turn be rewritten[1] as

(3) $\Box^n$(If E is true, so is $\Box^i$H)

(where the 'if' itself, if taken out of context, can safely be regarded as truth-functional). But, if (3) is a logically more primitive form of '$s[H, E] \geqslant i/n$', what is the corresponding analysans for '$s[H, E] \leqslant i/n$'? Obviously it is

(4) Not-$\Box^n$(If E is true, so is $\Box^j$H)

where $j = i + 1$. For we want to be able to say that the support given by E to H does not attain a higher grade than the ith. Moreover, since we

[1] It is here being assumed that the top grade of support assigned by higher-order assessments to lower-order support-assessments in a given field may be represented by the same symbol as the top-grade support that is sometimes assigned by these lower-order assessments. There are two reasons for such an assumption. The first is that it considerably facilitates the systematisation of inductive syntax, as emerges in §§21–22 below. The second is that all fully supported hypotheses have the same standing (cf. §10, p. 94), for if a hypothesis really has top grade support according to some appropriate support-function it cannot be more strongly established. It is also plausible and convenient to require that lower-order assessments should only be accepted if they are themselves thought to be fully supported (at whatever sacrifice of simplicity this involves: cf. p. 92, n. 1).

have in any case to distinguish between 's[H, E] $\leqslant i/n$' and 'If E is true, then s[H] $\leqslant i/n$' (see §8, p. 68 f. above), we can now formulate the latter as

(5) $\square^n$(If E is true, so is not-$\square^j$H)

where $j = i + 1$. The difference between (4) and (5) articulates the logical difference between E's not stating evidence of more than *i*th-grade support for H, and E's stating evidence of there being not more than *i*th grade support for H, respectively.

What these analyses imply will be investigated formally in the following chapter. Our immediate concern is the bearing they have on the problem of inductive ratification. It is now clear that the correct comparison to make, in our search for ratificatory analogies, is not between the all-or-nothing deducibility, and the varying inducibility, of one proposition from another. What varies in inductive inference is not the strength of the licence to move from premiss to conclusion, as it were, but rather the strength of the conclusion that we are licensed to derive from the premiss. So, specifically, the terms of our comparison are a logically warranted inference from E to H's being logically true, on the one side, and an inductively warranted inference from E to H's having so-or-so much inductive support, on the other side. It remains to be seen whether a satisfactorily extensive analogy is forthcoming on this basis of comparison; and we shall have to bear in mind that the analogy must now be forthcoming in two separate places. A logically warranted inference has to be compared with an inductively warranted one—i.e. with an inference warranted by a fully supported hypothesis—and at the same time a conclusion that a proposition is logically true has to be compared with a conclusion that it has so-or-so much inductive support.

It is obvious that this particular kind of deductive, or logically warranted, inference has at least one necessary-but-not-sufficient condition which it does not share with inductive inference when the latter is correspondingly conceived. If the premiss of the deductive inference implies (deductively) the logical truth of H it also implies *a fortiori* the truth of H. But if the premiss of the inductive inference implies less than full inductive support for H it does not imply (inductively) the truth of H.

The situation is much better however if the direction of comparison is reversed. The square-operator in (3) turns out to have a logical syntax that is a part of the logical syntax of 'it is logically true that', so that all the necessary-but-not-sufficient conditions of '$\square^i$ . . .' are also necessary-but-not-sufficient conditions of 'it is logically true that . . .'. For example the consequence and conjunction principles for hypotheses have obvious analogues, and so does the uniformity principle (cf. metatheorems 344–6, 351, 357, 445–447, 452 and 475 in §22, p. 228 ff. below). We again (cf. p. 196 above) have a situation in which the sense of 'valid reasoning', 'support', etc. would appear as if extended to cover induction by a process analogous to that by which, for example, the sense of 'number' has been extended to cover fractions, surds, etc.

Moreover, the equivalence of support-functions to simplification-functions (cf. §16) points to the existence of a certain spectrum, with logical truth at one extreme, and no support at all from test-results at the other. In order to see this point it is necessary to bear in mind that, if a proposition is inferred to be logically true, the proposition must be supposed to have been shown to hold under all uniform replacements of its non-logical terms. Now, non-logical terms, in the propositions we are concerned with here, are either individual constants or predicate constants. In regard to the former there is a complete analogy between propositions about logical truth and propositions about inductive support, since, as a consequence of the instantial conjunction principle —(4) of §2—, assessments of inductive support are invariant in truth-value under all uniform replacements of their individual constants. But what about predicate constants? If a member, U, of a particular class of materially similar universal hypotheses is inferred to have full suport, U must have been shown in effect to hold under all uniform simplifying-or-complicating replacements of its predicate terms—i.e. all inductively modified versions of U have full support. If U is inferred to have some support, then U has been shown to hold under some uniform simplifying-or-complicating replacement of its non-logical terms—i.e. some version of U has full support; and the higher the grade of support that U itself has the greater is the number of different grades of simplicity that fully supported versions of U can have. Finally, if U is inferred to have no support, it must have been shown not to hold under any uniform simplifying-or-complicating replacement except a tautologising one—i.e. only a 0th grade version of U has full support. (There is some similarity here to the spectrum in Carnap's theory that runs from the total inclusion of one range in another to the total exclusion. But the latter spectrum has logical implication and logical contradiction at its extremes, whereas the present one has logical truth at one extreme and the non-existence of positive support at the other.)

These analogies are capable of systematic deployment, because the logical syntax of inductive support, as developed in the preceding chapters, may be mapped *in toto* on to a class of generalisations of the Lewis-Barcan modal system S4, as will be shown in the next chapter. So far as we may assume, on the one hand, that S4's assertions about logical truth are correct, and, on the other, that inductive support is subject to all and only the syntactic principles already described *plus* their logical consequences, it is demonstrable that the necessary-but-not-sufficient—i.e. the syntactic—conditions of inductive support are also necessary-but-not-sufficient for logical truth.

How well substantiated, then is this double assumption? It will be shown informally in the following section (§21) that what the Lewis-Barcan system S4 asserts about logical truth is correct, if by calling a proposition logically true we mean that it is true under all uniform-replacements of its non-logical terms. Also, arguments have been given in preceding chapters to show the correctness of all the principles of inductive syntax that have been asserted. But it is difficult to conceive of a philosophical argument which could show that *only* those principles and their logical consequences are correct for the logical syntax of inductive support. This is not an issue that could be settled by a meta-mathematical proof of strong or weak completeness—quite apart from the well-known difficulties about completeness proofs for second-order quantification theory. It would be of no use to prove that none of its non-derivable formulas could consistently be added to the axioms of a particular formal system that has an appropriate interpretation, since such a proof would have to take the system's formation-rules as given; whereas what is at stake here, in relation to a particular formal system (under an appropriate interpretation), is whether all the principles of inductive syntax are both expressible in the system and also provable, not merely whether any that are expressible are provable. Nor could the issue be settled by a metamathematical proof of weak completeness, whereby on a given semantical definition of truth for well-formed formulae of the system it is shown that every true formula is a theorem. For by a proof that the formal system is weakly complete in relation to some chosen semantical model one would do nothing to show that the chosen model is itself complete, or comprehensive, in relation to the informal inductive discourse of scientists, lawyers, etc. It seems therefore that at this point we can only challenge objectors to produce their counter-examples. A putative counter-example would be a principle of inductive syntax that has no analogue for logical truth and applies equally to all types of inductive function (or at any rate to a wide variety of the functions already examined). But even if such a syntactic principle were in fact entrenched in ordinary discourse it might still be open to us to argue that its peculiar disanalogy with logical truth deprives it of any right or title to control human reasonings, rather than admit that this disanalogy constitutes a bar to the ratification of induction.

A more serious objection, however, is certainly possible. 'I am willing to grant,' an objector may say, 'that all the necessary-but-not-sufficient conditions of inductive support, as you conceive it, are also necessary-but-not-sufficient conditions of logical truth on the proposed basis of comparison. But is this an adequate legitimation for any reputedly valid inductive inference? You yourself objected to Carnap's system, as a basis for ratification, that the analogies with logical truth are equally good for any of his infinitely many symmetrical c-functions, so that there is nothing to discriminate between c^*, say, which allows addition of new instances to increase confirmation, and c^+, which does not. But may not precisely the same kind of objection be raised against using your own system of inductive syntax as a basis for ratification? So far as the principles asserted are purely syntactic they fit any inductive support-function whatever. It seems just as legitimate to call an inference valid if it is drawn from one favourable test-result as from another. No distinction is drawn between the manipulation of relevant variables and of irrelevant ones.'

The root of the trouble here, of course, is that our ratificatory analogies must perforce relate conditions that are merely necessary, and not also sufficient, for the validity of an inference, since we cannot expect sufficient conditions for induction to be either necessary or sufficient for deduction. Nevertheless there is one reason why the proposed analysis of induction is not as poor a basis for ratification as Carnap's system is. Whereas Carnap has to invoke *a priori* intuitions or assumptions about the nature of induction in order to justify selecting c^* (or adopting his preferred value for λ), the present analysis of inductive support entails that, when we come to consider features of inductive inference that can no longer be ratified by analogies with deduction, we can turn directly to empirical evidence about the relevance of varying one type of circumstance rather than another.

Admittedly, inference of this relevance from the evidence is itself a form of inductive inference. It involves the supposition that past instances of a certain variable's being relevant to a particular class of materially similar hypotheses are grounds for concluding that it has present and future relevance. Hence, if we apply the term 'valid reasoning' to such an inference about relevance, our use of this term here has again to be ratified. So that there comes a point, even on the present basis of comparison, where our quest for ultimate ratification appears to be unending, because it seems to lead into an infinite regress. It looks as though we must either give up the quest or commit the well-known fallacy of proposing to justify induction by an appeal to premisses that can themselves be supported only by induction.

But this is not the right way to look at the situation. To look at it thus assumes that there is a known, right method of assessing the support that stems from favourable results under different kinds of test, and that what we have to do is to justify using the term 'valid reasoning' in a sense that covers such a known, right method of inference. But there is no conclusively known, right method. Our criteria for assessing inductive support in natural science are no less open to revision and correction in the light of further experience than is any other scientific hypothesis that we believe to be fully supported, as the thalidomide case has shown. If the discovery of a hidden variable and of its hitherto undetected effects forces us to revise a hypothesis, it at the same time forces us to revise our criteria of inductive assessment for all materially similar hypotheses. The discovery that thalidomide was liable to produce toxic effects in pregnancy was at the same time a discovery affecting the kind of evidence from which we can arrive, by inductive inference, at a fully supported generalisation in certain fields of pharmacological enquiry. Hence our quest for an ultimate ratification of induction plunges us rather into an infinite progress, as it were, than into an infinite regress. The use of 'valid reasoning', and related terms, in the relevant sense, is partly justified by syntactic analogies between inductive inferences, on the one hand, and deductive inferences about what is logically true, on the other. But beyond this point the usage is not, and can never be, fully justified by syntactic analogies with deduction, because its criteria beyond this point are open to indefinite correction in the light of future experience and do not consist in the possession or non-possession of purely logical properties.

But if the quest for ratification is indeed blocked beyond here, this is not because the project was misconceived or fallacious.

For, first, from a certain point of view the analogy with deduction is maintained. Just as the rules of inference used in deductions about what is logically true, within a system of axiomatised modal logic, are deducible within some other deductive system, though not in that one, so too the criteria used in the appraisal of inductive support from given empirical evidence are themselves supported by empirical evidence, though not —for the most part—by the same evidence.

Secondly, in appraising the basis for extending the sense, and therewith the denotation, of a term we are bound to come across dissimilarities as well as similarities between what is denoted by the term in its old sense and what is denoted in the new: otherwise there would be no such extension. The eye of a needle is aptly so termed because it is an aperture shaped like a human eye, and no-one thinks it scandalous to term it thus because it is not also an organ of sight. Correspondingly it is an essential feature of inductive inference that beyond a certain point its criteria of validity are only justifiable by induction, and not at all by analogy with deduction; nor is there anything scandalous to philosophy about having to say so. It is enough to have shown that in respect of certain of its logical properties (viz. its necessary-but-not-sufficient conditions) the concept of inductive support is a generalisation of the concept of logical truth.

Thirdly, when we talk about this inductive justification of inductive criteria we are not asserting, like J. S. Mill, that belief in the uniformity of nature is inductively justifiable. We are not asserting the uniformity of nature in the sense that a true universal hypothesis of some unspecified kind must in principle be available for the prediction of any observable event. Nor are we claiming, like J. M. Keynes, that belief in the limited independent variety of nature is inductively justifiable. For such excessively general beliefs are quite untestable. No conceivable evidence could refute them. Nor are they needed for a ratification, as distinct from a validation, of induction. What we are instead asserting is that, in each field of scientific enquiry, beliefs about which variables are relevant are capable of no better, and no worse, than inductive support. Indeed, these beliefs are not only corrigible in principle, but also often modified in practice, whereas Mill and Keynes supposed their principles to have been constantly confirmed by experience.[1]

Fourthly, even the syntactic analogies with deduction suffice to make the ratification of induction safe against the main kinds of objection that have rightly been urged against attempts to vindicate induction.[2] Would-be vindications of induction tend to claim on its behalf that it is the only self-correcting policy for making predictions about the future. But philosophers have rightly objected that many other self-correcting policies for doing this are conceivable. For example, a counter-inductionist, who took past cases of R's being S's as evidence that future R's would not be S's, might argue that past failures of counter-induction should be taken as evidence for its future success. Nevertheless, the counter-inductionist's mode of prediction does not deserve the title of valid reasoning, if syntactic analogies with deducibility have to be preserved. What characterises his mode of prediction is that the only evidence to which he pays any attention is outside the domain of his conclusion: the conclusion is about all future events, and the evidence is in the past. Hence the existence of a counter-instance (in the future) to any generalisation (about the future) that he infers does not, for him, exclude the existence of full support for the generalisation. But this is contrary to one feature of the syntactic analogy between induction and

[1] J. S. Mill, *System of Logic*, III, xxi, 4, and J. M. Keynes, *A Treatise on Probability*, p. 260.
[2] Cf. M. Black, *Problems of Analysis* (1954), p. 169 ff.

the deduction of conclusions about logical truth. Such an analogy requires that a fully supported conditional, should always license inference from its antecedent to its consequent, as in (3); and it accordingly implies that 'H is fully supported' is contradicted, where H is a singular or generalised conditional, by the negation of H (cf. metatheorems 321 and 464 in §22 below). Of course, the counter-inductionist might try to combine respect for this feature of inferential syntax, in relation to the upper limit of evidential support, with insistence on his own criteria for the lower limit. I.e., the counter-inductionist might come to hold both that past cases of R's being S's give some support to the hypothesis that future R's will not be S's, and that this hypothesis does not have full support if some future R is an S. But then instead of taking his counter-inductive method to be fully supported by constant failure, he would have to accept, in advance, that any failure of the method excluded it from having full support. In short, he would cease to be a counter-inductionist.

What about a fortune-teller, then, who makes predictions about other people's lives from patterns of tea-leaves at the bottom of his own tea-cup? Such predictions might well conform to the logical syntax of (3), because the fortune-teller could take predictive failure as a ground for revising his method of interpreting tea-leaf patterns. The fortune-teller's procedure would thus deserve the title of 'valid reasoning' according to the present proposal. But why not? Certainly that procedure is not a standard interpretation for inductive syntax. It is not the same method of reasoning as that currently used in experimental science. Presumably, also, its criteria would require much more frequent revision than those of experimental science. But the fortune-teller would produce just the same conclusions as an experimental scientist who mistakenly believed that the patterns of his own tea-leaves had a causal effect on other people's lives and was constantly revising his hypotheses about the nature of this causal connection. Such a scientist would not be irrational in the structure of his argument but only in the restrictions he imposes on himself in the collection of evidence and in his disregard of already established hypotheses which imply severe limitations on the causal potential of tea-leaf patterns. His conclusions would admittedly be unacceptable, but this would be because of his unacceptably narrow premisses, not because of invalid reasoning.

Finally, it may be objected that what have been ratified are inferences from premisses about experimental test-results to conclusions about the grade of inductive support that exists for certain hypotheses, whereas the inferences with which Humeian sceptics have been typically concerned are inferences from those premisses to conclusions about the truth of these hypotheses. But though the decision to regard inductive support as a matter of degree, rather than of all-or-nothing, does indeed enforce this change in our conception of the problem, nevertheless from the conclusion that a given hypothesis is fully supported we can always infer the desired conclusion that it is true; and this latter type of inference, too, is licensed by the logical syntax of inductive support that is mapped, in the next chapter, on to a generalisation of the logical syntax of logical truth. So the transition of mind that philosophers have often referred to as an inductive inference is covered by the present account. Only it turns out to be a two-stage, not a one-stage, piece of reasoning. It embraces a move from the truth of E in (3), say, to the truth of $\Box^n$ H, and a move from the truth of $\Box^n$ H to the truth of H. Indeed, it is arguable, therefore, if we agree to regard inductive support as a matter of degree, that the underlying problem is more clearly formulated as a problem about the ratification of inductive support than about that of inductive inference. So the 'inference' terminology is avoided in most other parts of the present book, and is used in this chapter only in order to establish liaison with other philosophers' discussions of the same issue.

VII

A Formalisation of Inductive Syntax

§21. Some formal characteristics of an inductive calculus

Several important purposes can be achieved only by mapping the logical syntax of inductive support on to an appropriate formal system or systems. It will here be mapped on to a class of generalisations of the Lewis-Barcan system S4. Choice of S4, rather than of any other of Lewis's systems, is determined for us both by the principles of inductive syntax and also by the conception of logical truth that proved fruitful in the preceding section, §20. These factors also favour Barcan's development of modal quantification theory, if suitable restrictions are placed on the latter's interpretation.

A formalisation for the logical syntax of inductive functions is understood here to be constituted by a description of primitive notations, rules for generating well-formed formulas out of these notations, criteria for selecting certain well-formed formulas as theorems, and a partial assignment of meanings to well-formed formulas, such that all the principles of inductive syntax are theorem-schemata under the proposed interpretation, or effectively recognisable as schemata for abbreviations of theorems, and nothing is a theorem-schema that is inconsistent with any of those principles.

But what is the point of constructing such a system? The major difficulty in any enquiry into the logical syntax of induction is to defend the correctness of certain syntactic principles, not to explore their implications and inter-connections. Hence, no amount of ingenious mathematics can compensate, in a proposed solution of this problem, for inadequate philosophical argument. Nevertheless, just as the construction of formalised systems of alleged inductive logic, without elaborate philosophical elucidation and defence, is a pointless and somewhat grotesque activity, so too an attempt to philosophise about the structure of inductive reasoning hardly deserves to be taken seriously unless its conclusions are presented in some rigorously systematic form. Perhaps the reasons for this are worth setting out in a little detail, partly because some of those who practise formal systematisation in inductive logic[1]

[1] On the present view there cannot be an inductive logic in Carnap's sense, or even in Mill's, since assessments of support are logically contingent. It is the logical syntax of these assessments, rather than their truth-criteria, that should be called 'inductive logic'.

give the impression that they regard such systematisation as an end in itself, and partly because some of those who do not practise it seem correspondingly unaware of the further ends that it is capable of achieving. One can readily understand how in the early excitement about the Frege-Russell enterprise opinion sometimes tended to swing to one extreme or the other, as on a Wittgensteinian see-saw between the spirit of the *Tractatus* and the spirit of the *Investigations*. But there is no excuse now for not treating formal systematisation as the indispensable tool that it is, in certain fields of enquiry. It is not just a fetish to be worshipped or ridiculed according to one's direction of interest, educational background or country of origin.

First, in a field where so much is still controversial and intuitions are notoriously slippery and deceptive, it is obviously wrong to propose any fundamental principles of inductive syntax without at the same time presenting a fair range of their logical consequences. One must be in a position to know what one is implicitly committed to by accepting the proposed syntax.

Secondly, a good deal of the controversy that has raged about inductive logic seems due to philosophers' having been engaged in the analysis of different concepts under the same name. It is therefore essential that any proposed inductive principles should be proved consistent with one another, for there is no other guarantee of their capacity to define a single, homogeneous type of function. But consistency is only provable within the context of an adequately rigorous formalisation.

Thirdly, the principles of inductive syntax are themselves as much susceptible of mutual corroboration by consilience as are any other inductively supported generalisations. It has already been shown how and why inductive consilience takes place (§9), and it has been pointed out how philosophical argument in favour of a particular set of syntactic principles can itself have an inductive structure (p. 1 f. and p. 183 f.). But consilience of this nature can be achieved only by means of unification into a more powerful system.

Fourthly, we must concede that probabilistic theories of inductive support have at least the merit of connecting the relatively unfamiliar and controversial concept of a support-function (or so-called 'confirmation-function'), as they envisage it, with a calculus that articulates the structure of many other, well-accepted bodies of thought as well. Something similar is desirable for any non-probabilistic syntax of inductive support—especially if it is to be ratified by analogies with logical truth, as suggested in the previous section (§20). In fact, what is desired here can be provided by a certain kind of generalised modal logic. So far as their syntax is concerned propositions about the values of support-functions turn out to be reducible to, i.e. definable as abbreviations of, expressions containing nothing but the symbolism of such a logic. In this way an indispensable requirement for theoretical unification, and consilience, is satisfied: viz. the introduction of a more fundamental terminology (cf. §9, p. 87). Indeed, the ratification of induction, as envisaged in the preceding section (§20), is itself an instance of consilience, since it involves unifying the syntax of logical truth and the syntax of inductive support within the framework of a common terminology.

Fifthly, it would be a mistake to ignore altogether the psychological argument for systematisation, as distinct from the above four epistemological arguments. While the fundamental principles of inductive syntax were being established, in the six preceeding chapters, they were—of necessity—introduced somewhat discursively and in a way that makes a synoptic judgement about them difficult. But if this syntax is now to be formalised its main principles can be presented in concise array and in an orderly and readily comprehensible style.

There may well be other important purposes that formal systematisation can achieve here. But for the five above-mentioned it is indispensable. So what kind of formal system is appropriate? A generalised modal logic turns out to fit exactly.

It is already known that C. I. Lewis's systems of strict implication may be generalised.[1] In this way it is possible to represent not only Lewis's standard modality 'it is logically necessary that', but also any desired number of other modalities of certain kinds, such as 'it is analytically necessary that' or 'it is physically necessary that'. It is also possible, within such a system, to distinguish any desired number of different kinds of non-extensionality. E.g., contexts that resist the substitution of truth-functional equivalents for one another may not be equally resistant to the substitution of logical or analytic equivalents. Instead of including just a single type of modal operator in our primitive notation, as in Lewis's systems, we include any desired number of modal operators, each with its differentiating numerical superscript, viz. $\Box^1$, $\Box^2$,.... Lewis's axioms are then generalised correspondingly. For example, in the generalisation of Lewis's S4 we are able to prove $\Box^i A \rightarrow \Box^i \Box^i A$. Moreover in each system we can prove $\Box^j A \rightarrow \Box^i A$, where $j \geqslant i$, so that all the square-operators needed can be arranged in a hierarchy of subalternation.

[1] L. Jonathan Cohen, *The Diversity of Meaning*, 2nd ed. (1966), p. 237 ff. However, in that book and in this one square-operators are taken as primitive and diamond-operators defined in terms of them, whereas Lewis's modal primitive in S4, was a diamond-operator—i.e. an operator signifying 'it is logically possible that', not 'it is logically necessary that'.

But what is the lower limit of such a hierarchy? One possibility[1] is to allow any well-formed formula A to be rewritten $\Box^0 A$, so that from $\Box^j A \rightarrow \Box^i A$, where $j \geqslant i$, we can derive $\Box^j A \rightarrow A$. In effect this treats 'it is true that' as the null case of modal prefix. But we certainly do not need to cut off the descending order of modal operators at this point. Just as 'It is logically necessary that . . .' (where the blank is to be thought of as filled by some proposition) or 'It is analytic that . . .' asserts that '. . .' is of stronger standing than would be conveyed by the mere assertion of '. . .' or of 'It is true that . . .,' so too there are expressions capable of asserting that '. . .' may be of weaker standing than this. If we say, for instance, 'There is some support for its being true that . . .' we leave it open whether '. . .' is actually true or not. Thus below the lowest modal operator which is such that the proposition formed by prefixing it to A logically implies A we may still place any number of other operators. At the lower limit of this hierarchy we could now place an operator $\Box^0$, such that $\Box^0 A$ is no longer a suitable way of rewriting A but is available for a much weaker interpretation, e.g. 'There is at least zero support for A's being true'. But 'it is true that' will no longer have any position in the hierarchy. We shall just have two designated superscripts, d and e, where $d > e > 0$, such that $\Box^d A$ is the strongest modalisation of A, $\Box^e A$ is the weakest modalisation of A that still implies the truth of A, and $\Box^0 A$ is the vanishing point, as it were, of A-assertion. Specifically, if there are n canonical tests, we should have $e = n$. I.e. d can denote the designated level of logical truth, e the level of establishment, or full inductive support, and 0 the level of zero-grade support. Moreover, since at least zero support will exist for A, whatever A says, $\Box^0 A$ should now be provable in our formal system. For we can take $\Box^0 A$ to be an abbreviation for the proposition that either there is at least some support for A or there is none. That is, where $\Box^x$ is an operator-variable taking primitive square-operators as its substitution-instances, $\Box^0 A$ is now to be treated as an abbreviation for the tautology $(\exists \Box^x) \Box^x A \vee -(\exists \Box^x) \Box^x A$.

It might be thought that instead of assigning the intermediate position to $\Box^e A$, because $\Box^e A$ still implies the truth of A, we should economise by allowing A to be rewritten as $\Box^e A$—thus assigning the intermediate position to A itself. But the trouble is that since we can prove $\Box^j A \rightarrow \Box^i A$, where $j > i$, we shall then also be able to prove $A \rightarrow \Box^i A$, where $i < e$, whereas it is not at all clear that the mere truth of A implies the existence of any inductive support for A. For example, it is very convenient to be able to construe universal hypotheses as universally quantified truth-functional conditionals, and this commits us to accepting that innumerably many universal hypotheses are true just because their antecedent

[1] Adopted ibid.

clause is never satisfied. But we do not want to be committed also to the principle that each such vacuously true hypothesis has some inductive support. Experimental science acknowledges no support for the hypothesis that, say, any perpetual-motion machine that exists is made of gold. Accordingly,[1] if we are to extend our hierarchy of modal operators to lower levels, at which $\Box^i A$ no longer implies A, we must regard 'it is true that' as being quite outside the hierarchy. We shall need to be able to prove $\Box^e A \rightarrow A$ but this will not be derivable from $\Box^j A \rightarrow \Box^i A$ where $j \geqslant i$.

There is also another thesis for which special provision needs to be made, if the hierarchy of square-operators is to be extended downwards. If A is logically true, we shall want it to be provable that there is no support for –A. Now where $i \geqslant e$ we shall have $\Box^i A \rightarrow A$, and it is easy enough to show therefrom (by contraposition and detachment) that if we can prove A we can also prove $-\Box^i -A$. But, since we are not to be able to prove $\Box^i A \rightarrow A$ where $i < e$, we cannot show therefrom that if A is provable so is $-\Box^i -A$. We need therefore to make special provision in our criteria of theoremhood to ensure that if A is provable so is $-\Box^i -A$, where $i < e$.

Lewis developed several modal systems, and other logicians have added to the list. But the system to be generalised here is determined for us by the conception of logical truth that proved particularly fruitful of ratificatory analogy in the preceding section (§20). We need to generalise the formal system which fits in best with the informally supported requirement that a proposition should be conceived to be logically true if and only if it is true under all uniform replacements of its non-logical terms.

Now, presumably 'it is logically true that' is not a non-logical term, and any proposition formed by prefixing this operator to another

[1] This is not, however, the only argument for excluding the provability of $A \rightarrow \Box^i A$ where $i < e$. If it were, we might think it worth while to abandon the truth-functional analysis for universal hypotheses, and adopt instead some analysis that does not allow any of them to be vacuously true. Another reason for not wanting to prove $A \rightarrow \Box^i A$ where $i < e$ is that with the help of metatheorem 228 (*q.v.*) in §22 below we could then prove $A \leftrightarrow \Box^i A$ where $i < e$, which would make use of $\Box^i$ quite pointless where $i < e$. Indeed, if we wanted to develop a syntax for support-functions in which the truth of 'A' implies the existence of support for 'A', the easiest way to do so would be not to prolong the hierarchy of square-operators below $\Box^e$, but instead to use diamond-operators, in reverse order, to express support—where $\Diamond^i A$ abbreviates $-\Box^i -A$. For $A \rightarrow \Diamond^i A$ would then be provable while $\Diamond^i A \rightarrow A$ would not. But the general conjunction principle would also not be provable and there could be positive support for each of two inconsistent hypotheses. Even if a critic were happy with this (despite the arguments of §8 above), he might be disappointed when he found that there was also then some inductive support for any proposition of which the negation was not a truth of logic—e.g. for 'Any perpetual-motion machine that exists is made of gold'.

proposition is presumably true if and only if that other proposition is logically true. It follows that if it is logically true that A, it is logically true that it is logically true that A. Hence the modal system we need must at least include the characteristic thesis of Lewis's S4. But it cannot include the characteristic thesis of S5, since $-\Box-\Box A \rightarrow \Box A$ should not be provable. Take A to represent any contingent proposition that can be transformed into a logical truth by some uniform replacement of its non-logical terms, such as 'If to-day is Tuesday, to-day is market-day'. Then such a proposition is not logically true, in that uniform replacement of its non-logical terms does not in every case yield a true proposition, while the denial that it is logically true is also not logically true. I.e. what we have is $-\Box A \mathbin{\&} -\Box-\Box A$, and so $-\Box-\Box A \rightarrow \Box A$ should not be provable. Nor should the latter thesis be disprovable, since what $-\Box A \mathbin{\&} -\Box-\Box A$ represents is not itself logically true. So the system we need to generalise is S4, not S5. (See also s.v. metatheorem 230 in §22.)

But what kind of quantification theory should go with S4? In non-modal quantification theory we have $A \rightarrow (x)A$ as an axiom or theorem only where x is not free in A—for the obvious reason that where x is free in A the closure of $A \rightarrow (x)A$ would appear to authorise generalisation from a single instance. It would appear to authorise inferring, say 'Anyone is wise' from 'Socrates is wise'. If however A is an axiom or theorem of our logic, we are rightly allowed in non-modal quantification theory to infer from A to the closure of A, i.e. to $(x)A$, since truths of logic are supposed to remain true under all uniform replacements of their non-logical terms. Hence we may regard it as a truth of logic that if A is a truth of logic then $x(A)$ is true. It follows that if $\Box A$ is to represent 'It is logically true that A', and all truths of logic are to be represented by axioms or theorems in modal quantification theory, then $\Box A \rightarrow (x)A$ ought to be provable in modal quantification theory, even where x is free in A. Now, in S4 if C is provable so is $\Box C$. Hence in S4's quantification theory we should be able to prove $\Box(\Box A \rightarrow (x)A)$. Moreover, since $\Box(\Box A \rightarrow B) \rightarrow (\Box A \rightarrow \Box B)$ is provable in S4, we should also be able to prove, by detachment, $\Box A \rightarrow \Box(x)A$; and then, by standard rules of quantification, we can derive the formula $(x)\Box A \rightarrow \Box(x)A$. Modal systems in which the latter formula is not derivable[1] are inadequate as

[1] E. J. Lemmon (abstract of paper entitled 'Quantified S4 and the Barcan formula', *Jour. Symb. Log.* xxv, 1960, p. 391 f.) proved that this formula is not a theorem of a system consisting of propositional S4 plus standard axioms of non-modal quantification theory. Such a proof, however, does not settle the question whether a quantified version of S4 should add nothing to the axioms of propositional S4 besides the standard axioms of non-modal quantification theory. Now obviously, if the intended interpretation for $\Box$ is 'it is provable that', then, for any omega-incomplete system of proof, $(x)(\Box A) \rightarrow \Box(x)A$ should not be provable in S4. Indeed, S. A. Kripke ('Semantical Considerations

formalisations of logical truth, in the sense of truth under all uniform replacements of non-logical terms. Correspondingly, in the system to be constructed here $\Box^{i}A \rightarrow (x)A$ ought to be provable where $i \geqslant e$. But we do not want to be able to prove quite so much where $i < e$, since this would entitle us to infer the truth of A from even the very slightest support for A. Where $i < e$ we shall just need $\Box^{i}A \rightarrow \Box^{i}(x)A$, in order to be able to prove the uniformity principle—(23) of §4.

We must be careful, however, to place appropriate restrictions on the interpretation of any modal system that contains such theses. For example, if a predicate symbol S may represent 'is identical with Socrates', and some individual constant denotes Socrates, we can apparently infer from the closure of $\Box Sx \rightarrow (x)Sx$, by universal instantiation and detachment, that anyone is identical with Socrates. It follows that, if individual constants, or unbound individual symbols, represent proper names, then no proper name may be part of what is represented by an unbound predicate letter. Or if the elements of our universe of discourse are identified by spatio-temporal references, then the terminology of such references must be similarly unrelated to what is represented by an unbound predicate letter. We do not need to ban certain types of predicate-meaning altogether, e.g. Goodman's 'grue'. But the interpretations permissible for predicate symbols are to be interdependent with, because wholly exclusive of, the terminology permissible for referring to elements in the domain of first-order quantification. We shall thus maintain the catholicity of our uniformity principle as a principle of logical syntax, while admitting its capacity for parochialism when it is semantically reinforced (see §18, p. 172 f.).

These restrictions on the interpretation of our formal symbols may be regarded as just one instance of a general type of restriction which is necessary in formal logic in order to avoid generating paradox. The general law is that each syntactically differentiated kind of symbol-occurrence in a formal system should be reserved, in the intended interpretation, for the performance of just one linguistic function, and just one kind of symbol-occurrence for each function. In other words, the structure of the interpreted system is to mirror the logico-linguistic structure of the informal discourse under analysis. No-one should be surprised if

in Modal Logic', *Acta Philosophica Fennica* xvi, 1963, p. 83 ff.) has shown that on a particular choice of model, or kind of model, this formula should not be provable even in a quantified version of S5. But if $\Box$ is to be interpreted as 'it is logically true that', in the sense proposed, there is equally no doubt—for the reason given in the text—that a quantified version of S4 should include the formula as a theorem (as in R. C. Barcan, 'A Functional Calculus of First Order Based on Strict Implication', *Jour. Symb. Log.* xi, 1946, p. 1 ff.). For a thorough discussion of this issue cf. G. E. Hughes and M. J. Cresswell, *An Introduction to Modal Logic* (1968), p. 142 ff., and esp. p. 182.

paradox sometimes results when the structure of an interpreted formal system does not mirror the structure of the discourse it purports to analyse.

Hence unbound occurrences of individual symbols are to function just as references to the individuals that constitute the domain of the informal discourse to be analysed; and unbound occurrences of first-order predicate symbols are to function just as expressions signifying kinds, circumstances or *n*-adic characteristics of these individuals. Accordingly no individual symbol may denote a symbol or formula of the calculus itself, and no interpretation of a predicate symbol may include a reference, direct or indirect, to such a symbol or formula. Also, no occurrence of a first-order predicate symbol may contain, in its interpretation, a reference, direct or indirect, to an individual. Similarly, unbound occurrences of second-order predicate symbols are not to signify anything but kinds, circumstances, or *n*-adic characteristics of what first-order predicate symbols signify, and no occurrence of a second-order predicate symbol is to contain in its interpretation a reference, direct or indirect, to anything signified by a first-order predicate symbol or individual symbol. In this way we may reasonably hope to avoid semantical antinomies like Grelling's paradox or the Liar. It is also in keeping with the general law proposed that, in order to avoid familiar paradoxes about quantification into non-extensional contexts, we must require expressions generating non-extensionality, like 'it is logically true that', to be represented only by the modal operator symbols and not to form any part of expressions represented by predicate symbols.[1] Nor may any predicate symbol represent assertions of existence: only the existential quantifier may do this. Otherwise a counter-example may arise[2] to $\Box^{i}(x)A \rightarrow (x)\Box^{i}A$, which should be provable if a theorem is to be obtained that represents the uniformity principle. Finally, though unbound individual symbols are to function as individual constants, in the intended interpretation, we do not need unbound occurrences of modal operator-variables to function as modal operator constants since the primitive square-operators are to function as those. It follows that unbound occurrences of modal operator-variables cannot be assigned a definite meaning in an interpreted calculus of inductive support, and are otiose. We may therefore regard formulas in which they occur as being not well-formed.

When we come to second-order quantification, however, we shall not want to be able to prove any such formula as $\Box^{i}A \rightarrow \Box^{i}(R)A$, where R is a predicate symbol that has one or more free occurrences in A, but only

[1] Cf. L. Jonathan Cohen, *The Diversity of Meaning*, 2nd ed. (1966), p. 233 ff.

[2] Cf. S. A. Kripke, op. cit., p. 90.

$\Box^{d}A \rightarrow (R)A$. For it emerged in the preceding section (§20) that invariance of truth-value under all uniform replacements of predicate-terms, though characteristic of logically true propositions, is not a syntactic feature of inductive assessments: rather, within certain limits, grade of support itself varies with the extent to which such replacements may be made. Correspondingly, while $\Box^{i}A \rightarrow \Box^{i}(x)A$ is needed as one of the premises from which the instantial comparability principle is proved (cf. metatheorems 358 and 476 below), $\Box^{i}A \rightarrow \Box(R)^{i}A$ is not needed as such a premise because this principle does not apply to higher than first-order generalisations (for the reasons described in §9 above).

In the informal analysis it was argued (in §9) that scientific theories had to be construed as involving third-order quantification. However, the logical syntax of inductive support for such hypotheses turned out to be substantially the same as that for hypotheses that involve only second-order quantification. Accordingly, in order to avoid unnecessary complications, the calculus that follows will admit only first- and second-order quantification. Similarly in the discussions of inductive support for statistical generalisations (in §13) hypotheses were conceived to be formulated in terms of sets, rather than in terms of kinds, characteristics or circumstances. However, to introduce class-abstraction operators, with appropriate formation-rules, definition-schemata, and criteria of theoremhood, would also be an unnecessary complication, since the logical syntax of inductive support for statistical hypotheses is not substantially different from that of inductive support for non-statistical hypotheses. Accordingly, the calculus that follows is developed only in terms of predicates, not in terms of set-names.[1]

One last point. It would be very arbitrary to assume that the number of relevant variables, and therewith the number of different canonical tests, is precisely the same in each field of experimental enquiry. A support-function appropriate to one field might allow for six grades of support, say, and one appropriate to another might allow for ten. It would follow that in the formal system that underlay the former—if $s[H] \geqslant e/e$ is represented in the system by $\Box^{e}$, as suggested in §20, p. 198 —we should have $e = 6$, and in the system that underlay the latter $e = 10$. We are therefore concerned here not with just a single formal

[1] For a treatment of class-abstraction operators in a generalisation of S4 cf. L. Jonathan Cohen, *The Diversity of Meaning*, 2nd ed. (1966) p. 245 f. Since I am excluding any connections between the terminology for referring to individuals and the terminology for expressing predicates, no first-order atomic proposition in the proposed type of interpretation can be other than contingent. Hence no modality *de re* is in point, and contingent identity is possible: cf. G. E. Hughes and M. J. Cresswell, *An Introduction to Modal Logic* (1968) p. 195 ff., esp. n. 151.

system, but rather with a class of such systems. Each system is open to alternative interpretations, in accordance with whether interest is focussed on inductive support, say, or on inductive simplification, or on one particular set of materially similar hypotheses rather than on another that has the same number of relevant variables. But the systems differ from one another in the values of e and d, and are thus suited to inductive functions that have differently sized ranges of values. Hence a general representation of the logical syntax of inductive support must describe what is common to all these formal systems, without restriction to particular values for e and d, and this is clearly best achieved by presenting the systems in a syntactical metalanguage which can refer indifferently to $\Box^e$ and $\Box^d$, rather than specifically to '$\Box^6$', '$\Box^{10}$', etc.

§22. A system of logical syntax for the grading of experimental support

A class of generalisations of quantified S4 is described, in a syntactical metalanguage, and an interpretation is proposed for them that makes the axiom-schemata represent previously discussed principles of modal or inductive logic. Monadic and dyadic inductive functors, and an inductive information-functor, are contextually defined in terms of the primitive symbolism. Metatheorems are developed in which the logical properties of the primitive modal operators are explored, and further metatheorems then explore the logical properties of the defined functors. These are elucidated, where appropriate, by reference to previous sections, and some comparisons are drawn with the properties of Carnap's symmetrical c-functions. It is shown that each of the defined functors satisfies standard requirements for expressions denoting the members of a well-ordered set. Such a system is demonstrably consistent if second-order non-modal quantification-theory is consistent.

The class of inductive calculi will be described in a metalanguage by means of symbols and strings of symbols which represent indifferently any corresponding symbols or formulas of any one such calculus.

Primitive symbols of an inductive calculus:—

An unbounded number of individual symbols, represented in the metalanguage by 'x', 'y' or 'z'.

An unbounded number of monadic, dyadic, etc. first-order predicate symbols, represented by 'R' or 'S'.

An unbounded number of monadic, dyadic, etc. second-order predicate symbols, represented by 'f' or 'g'. (It would be possible to add predicate symbols of yet higher orders, so as to cover, e.g., scientific theories that generalise about the various correlations which hold in a particular field of experimental research. But cf. §21, p. 215.)

A designated number d of modal operator-constants, viz. squares differentiated by superscripts that denote the positive integers 1, 2, 3, ..., d. '$\Box^d$' represents the square with the superscript that denotes the designated integer. '$\Box^e$' represents some particular square with a superscript denoting a number that is less than d, and 'e' refers to this number. '$\Box^i$' and '$\Box^j$' represent any primitive modal operator-constant indifferently, unless otherwise stated, and 'i' and 'j', respectively, refer to the numbers denoted by the superscripts of the squares so represented.

An unbounded number of modal operator-variables, represented by '$\Box^x$' or '$\Box^y$'.

Brackets, dash, ampersand and quantifier, represented by '( )', '–', '&', and '∃', respectively.

Individual and predicate symbols, and modal variables, are also represented indifferently, unless otherwise specified, by 'u' and 'w'; and numerical subscripts will be used to generate additional metalinguistic symbols where necessary.

A *well-formed formula* (wff) is defined recursively:—

$Rx_1x_2 \ldots x_n$ is a wff where R is n-adic. $f(R_1R_2 \ldots R_n)$ is a wff where f is n-adic. If A and B are wff, so are $-A$, (A & B), $\Box^i$A, and $(\exists u)$A. If A is a wff, and B differs from A only in having an occurrence of $\Box^x$ in one or more places in which A has an occurrence of $\Box^i$, then $(\exists\Box^x)$B is a wff. No other combination of the primitive symbols is well-formed.

An occurrence of u in A is a *bound* occurrence if in a well-formed part of A of the form $(\exists u)$B. Otherwise it is *free*. It follows that no occurrence of $\Box^x$ in a wff is free.

Hereafter the upper-case letters E, F, G, H, I, and J represent wff.

The *interpretation* envisaged for wff of such a calculus rests on the usual formal-logical replacements for brackets (punctuation), dash ('it is not the case that'), and ampersand ('it is the case both that ... and that ---'). Some restrictions on the replacement of free occurrences of predicate symbols and free occurrences of individual symbols have been mentioned in the preceding section (§21). Second-order predicate symbols will suffice, for present purposes, not only (when monadic) to signify membership of a particular natural variable, as in the antecedent of (7) of §9, but also (when dyadic) to signify a particular function that maps one first-order predicate on to another, as in the consequent of (7) of §9. In interpreting a wff each occurrence of $Rx_1x_2 \ldots x_n$ is to be replaced by '$x_1, x_2, \ldots x_n$ instantiate(s) R', and each occurrence of $f(R_1 R_2 \ldots R_n)$ by '$R_1, R_2, \ldots R_n$ instantiate(s) f'. The initial $(\exists x)$ in each well-formed part $(\exists x)$H of a wff, thus partially interpreted, is to be replaced by 'at least one thing is such that'; free occurrences of x, in such an occurrence of H, are to be replaced by occurrences of a pronoun grammatically connected with this occurrence of 'thing'; and 'at least one thing is such that...', where '...' stands for the sentence replacing H here, is understood to be true if and only if there is at least one individual symbol y and at least one element in the domain of discourse, such that free occurrences of y denote this element and a true proposition is got from '...' by putting y for each of the above-mentioned occurrences of a pronoun in '...': it is assumed that if y has a bound occurrence in $(\exists x)$H then each occurrence of y in the wff to be interpreted has been replaced by occurrences of some other individual symbol, which does not occur elsewhere in the formula. The initial $(\exists R)$ in each well-formed part of $(\exists R)$H of a wff is to be replaced by 'at least one kind, property or characteristic is such that'; and free occurrences of R in such an occurrence of H, are to be replaced analogously to free occurrences of x above. $\Box^{d}$ in $\Box^{d}$ H is to be replaced by 'it is logically true that'. $\Box^{i}$ in $\Box^{i}$ H, where $d > i > e$, may be replaced by any appropriate interpretation, such as 'it is analytically true that'. $\Box^{i}$ in $\Box^{i}$ H, where $e \geqslant i$, is to be replaced by 'there is at least ith grade inductive support (out of the total of e grades that are possible according to the criteria appropriate for ...-type hypotheses) for the proposition that', or by a corresponding non-standard interpretation, where e different canonical tests are possible for ...-type hypotheses. Grades of inductive support are determined in the following way, so far as the antecedent of a testable hypothesis mentions only one variable or only one variant of a variable:

(i) Some semantical category of non-logical expressions is used to determine (p. 39 ff.) the unmodified members (p. 146 f.) of a particular type of materially similar hypotheses.

(ii) Some natural (or legal, or moral, etc.) variables not expressible in that terminology are taken, as a matter of contingent fact, to be inductively relevant (p. 36 ff.) for that class of hypotheses.

(iii) These relevant variables, or an appropriate sub-set of them (p. 38), are listed in the order determined by some acceptable well-ordering relation (p. 57 ff.).

(iv) For any testable (p. 96 ff.) member, U, of that particular class of materially similar hypotheses, a hierarchy of canonical tests—of cumulatively increasing thoroughness—is determined (p. 53 ff.) from the ordered list of relevant variables preceded by an appropriately selected sub-set of the variable that is mentioned, or has a variant that is mentioned, in the antecedent of U (p. 80).

(v) Support for a testable proposition is graded in accordance with the results of canonical tests on it (p. 60 f.), whether these are contrived or occur naturally; and support for a non-testable proposition is graded either in accordance with its non-contingent status (p. 149) or, if it is contingent, in accordance with any support that exists for any one or more testable propositions to which it is both materially similar and also logically related in certain ways (p. 99).

Finally, the initial $(\exists\square^x)$ in each well-formed part $(\exists\square^x)A$ of a wff is to be replaced by 'there is at least some grade of support such that' and free occurrences of $\square^x$ in such an occurrence of A by 'there is this grade of support for the proposition that', where the expression 'this grade of support' refers back to the initially mentioned grade.

Definition-schemata are metalinguistic statements that describe permissible alternative ways of writing wff and well-formed parts of wff. In each case the new metalinguistic symbolism used should be taken to represent symbolism of similar shape to itself. Only definition-schemata 1, 2, 5, and 7 are used in stating the criteria of theoremhood. The others are used at various stages in the development of the metatheorems and are collected here for convenience of reference.

1. $(H \rightarrow I)$ for $-(H \,\&\, -I)$.
2. $(u)H$ for $-(\exists u)-H$.
3. $(H \leftrightarrow I)$ for $((H \rightarrow I) \,\&\, (I \rightarrow H))$.
4. $(H \vee I)$ for $(-H \rightarrow I)$.
5. $(H \rightarrow^i I)$ for $\square^i(H \rightarrow I)$.
6. $(H \leftrightarrow^i I)$ for $((H \rightarrow^i I) \,\&\, (I \rightarrow^i H))$.
7. $\diamond^i H$ for $-\square^i -H$.
8. $\square^0 H$ for $(\exists\square^x)\square^x H \vee -(\exists\square^x)\square^x H$.

Hereafter '$\square^i$' and '$\square^j$' represent any primitive modal operator-constant, or $\square^0$, indifferently, unless otherwise stated.

The next eight definition-schemata (9–15) introduce contextually a monadic (inductive) functor s[ ], the logical syntax of which is deployed in metatheorems 301–360 and 482–499. It is to be noted that according to these definitions the expressions s[ ] $\geqslant i/e$, etc., perform the linguistic role of proposition-forming operators on propositions, whereas in preceding chapters assessments of support for a proposition have been treated as propositions about propositions. The latter is certainly the natural mode of speech for informal treatment, and is adopted by Carnap in his formalised confirmation-theory also. But, while Carnap's object-language is thus wholly truth-functional, the price he pays for avoiding use of some kind of modal calculus is that his confirmation-functions are defined in an unformalised semantical meta-language. Here, instead, the expressions s[H] $\geqslant i/e$, etc. are defined in the object language, and so their logic is fully formalised. At the same time use of a modal calculus makes possible a short and simple formulation of the ratificatory analogies with logical truth. It may well be theoretically possible to deploy these analogies, and systematise the whole logical syntax of inductive support-functions, in a formalised semantical metalanguage, or metametalanguage, that is wholly truth-functional. But only an irrational prejudice against modal logics, engendered by the paradox-ridden or mystery-mongering interpretations that are (unnecessarily) sometimes given them, could make the complexities of such an undertaking appear worthwhile.

9. $\mathrm{s}[\mathrm{H}] \geqslant i/e$ for $\Box^{\mathrm{i}}\,\mathrm{H}$, where $e \geqslant i \geqslant 0$.
10. $\mathrm{s}[\mathrm{H}] < i/e$ for $-\mathrm{s}[\mathrm{H}] \geqslant i/e$, where $e \geqslant i \geqslant 0$.
11. $\mathrm{s}[\mathrm{H}] \leqslant i/e$ for $-\Box^{\mathrm{j}}\,\mathrm{H}$, where $e \geqslant i = (j-1) \geqslant 0$.
12. $\mathrm{s}[\mathrm{H}] > i/e$ for $-\mathrm{s}[\mathrm{H}] \leqslant i/e$, where $e \geqslant i \geqslant 0$.
13. $\mathrm{s}[\mathrm{H}] = i/e$ for $(\mathrm{s}[\mathrm{H}] \geqslant i/e \;\&\; \mathrm{s}[\mathrm{H}] \leqslant i/e)$, where $e \geqslant i \geqslant 0$.
14. $\mathrm{s}[\mathrm{H}] \geqslant \mathrm{s}[\mathrm{I}]$ for $(\Box^{\mathrm{x}})(\Box^{\mathrm{x}}\,\mathrm{I} \rightarrow \Box^{\mathrm{x}}\,\mathrm{H})$.
15. $\mathrm{s}[\mathrm{H}] > \mathrm{s}[\mathrm{I}]$ for $-\mathrm{s}[\mathrm{I}] \geqslant \mathrm{s}[\mathrm{H}]$.
16. $\mathrm{s}[\mathrm{H}] = \mathrm{s}[\mathrm{I}]$ for $(\mathrm{s}[\mathrm{H}] \geqslant \mathrm{s}[\mathrm{I}] \;\&\; \mathrm{s}[\mathrm{I}] \geqslant \mathrm{s}[\mathrm{H}])$.

The next eight definition-schemata (17–24) introduce contextually the corresponding dyadic (inductive) functor s[,], the logical syntax of which is deployed in metatheorems 401–499.

17. $\mathrm{s}[\mathrm{H}, \mathrm{E}] \geqslant i/e$ for $(\mathrm{E} \rightarrow^{\mathrm{e}} \Box^{\mathrm{i}}\,\mathrm{H})$ where $e \geqslant i \geqslant 0$.
18. $\mathrm{s}[\mathrm{H}, \mathrm{E}] < i/e$ for $-\mathrm{s}[\mathrm{H}, \mathrm{E}] \geqslant i/e$ where $e \geqslant i \geqslant 0$.
19. $\mathrm{s}[\mathrm{H}, \mathrm{E}] \leqslant i/e$ for $-(\mathrm{E} \rightarrow^{\mathrm{e}} \Box^{\mathrm{j}}\,\mathrm{H})$, where $e \geqslant i = (j-1) \geqslant 0$.[1]
20. $\mathrm{s}[\mathrm{H}, \mathrm{E}] > i/e$ for $-\mathrm{s}[\mathrm{H}, \mathrm{E}] \leqslant i/e$ where $e \geqslant i \geqslant 0$.
21. $\mathrm{s}[\mathrm{H}, \mathrm{E}] = i/e$ for $(\mathrm{s}[\mathrm{H}, \mathrm{E}] \geqslant i/e \;\&\; \mathrm{s}[\mathrm{H}, \mathrm{E}] \leqslant i/e)$ where $e \geqslant i \geqslant 0$.
22. $\mathrm{s}[\mathrm{H}, \mathrm{E}] \geqslant \mathrm{s}[\mathrm{I}, \mathrm{F}]$ for $(\Box^{\mathrm{x}})((\mathrm{F} \rightarrow^{\mathrm{e}} \Box^{\mathrm{x}}\,\mathrm{I}) \rightarrow (\mathrm{E} \rightarrow^{\mathrm{e}} \Box^{\mathrm{x}}\,\mathrm{H}))$.
23. $\mathrm{s}(\mathrm{H}, \mathrm{E}] > \mathrm{s}[\mathrm{I}, \mathrm{F}]$ for $-\mathrm{s}[\mathrm{I}, \mathrm{F}] \geqslant \mathrm{s}[\mathrm{H}, \mathrm{E}]$.
24. $\mathrm{s}[\mathrm{H}, \mathrm{E}] = \mathrm{s}[\mathrm{I}, \mathrm{F}]$ for $(\mathrm{s}[\mathrm{H}, \mathrm{E}] \geqslant \mathrm{s}[\mathrm{I}, \mathrm{F}] \;\&\; \mathrm{s}[\mathrm{I}, \mathrm{F}] \geqslant \mathrm{s}[\mathrm{H}, \mathrm{E}])$.

Finally, eight definition-schemata (25–32) are proposed for a dyadic factor i[,], that is interpretable, in i[E, H], as denoting the level of inductive information that is supplied by E, in relation to the question

[1] In an earlier version of this calculus (*Brit. Jour. Phil. Sci.* xvii, 1966, p. 105 ff., as amended in xix, 1968, p. 71) I defined $\mathrm{s}[\mathrm{H},\mathrm{E}] \leqslant i/e$ as, in effect, $(\Box^{\mathrm{x}})((\mathrm{E} \rightarrow^{\mathrm{e}} \Box^{\mathrm{x}}\,\mathrm{H}) \rightarrow (p)(\Box^{\mathrm{i}} p \rightarrow^{\mathrm{d}} \Box^{\mathrm{x}} p))$, where '$p$' represents a propositional variable. This definition is compatible with having a non-denumerable infinity of primitive modal operator-constants, whereas definition-schemata 11 and 19 are not. But, for the reasons given in Chapter II above, inductive functions should not be attributed a continuum of values. Moreover, besides requiring us to introduce the usual axioms for propositional variables, this definition would also have the disadvantage of forcing us to include an existence-postulate $(\exists p) \Diamond^{\mathrm{d}}(\Box^{\mathrm{i}}\, p \;\&\; -\Box^{\mathrm{j}}\, p)$, where $j > i$, if we wished to be able to prove $\mathrm{s}[\mathrm{H},\mathrm{E}] \geqslant j \rightarrow \mathrm{s}[\mathrm{H},\mathrm{E}] > i$ where $j > i$ (metatheorem 430 below).

whether H is true (cf. §15). The logical syntax of this functor is deployed in metatheorems 501–523 below.

25. $i[E, H] \geqslant i/e$ for $(s[H, E] \geqslant i/e \lor s[-H, E] \geqslant i/e)$.
26. $i[E, H] < i/e$ for $-i[E, H] \geqslant i/e$.
27. $i[E, H] \leqslant i/e$ for $(s[H, E] \leqslant i/e \,\&\, s[-H, E] \leqslant i/e)$.
28. $i[E, H] > i/e$ for $-i[E, H] \leqslant i/e$.
29. $i[E, H] = i/e$ for $(i[E, H] \geqslant i/e \,\&\, i[E, H] \leqslant i/e)$.
30. $i[E, H] \geqslant i[F, I]$ for $(\Box^x)(((F \rightarrow^e \Box^x I) \lor (F \rightarrow^e \Box^x -I)) \rightarrow ((E \rightarrow^e \Box^x H) \lor (E \rightarrow^e \Box^x -H)))$.
31. $i[E, H] > i[F, I]$ for $-i[F, I] \geqslant i[E, H]$.
32. $i[E, H] = i[F, I]$ for $(i[E, H] \geqslant i[F, I] \,\&\, i[F, I] \geqslant i[E, H])$.

Hereafter i/e will be further abbreviated to i; outermost brackets will always be omitted; and occasionally other brackets will also be omitted where their replacement is univocally determined by the formation-rules and definition-schemata.

A *metatheorem* is an assertion that has the form 'H is provable', which may be abbreviated to '$\vdash$H'.

The primitive *criteria of theoremhood* are as follows:– (An informal argument for 104 was given in §21 (p. 211); for 105 and 106, in §8 (p. 70 f.); for the serial ordering prescribed by 107, in §7 (p. 52 ff.); for 108, in §10 (p. 92, n.1); and for 109 and 110 in §21 (p. 212 ff.). Criteria 101, 102, 103, 105 and 107 are needed for the generalisation of S4 in respect of its modal operator, as has been shown elsewhere;[1] and 111–116 are appropriate adaptations of standard quantification principles.)

101. If H is a truth-functional tautology, then $\vdash$H.
102. If $\vdash$H and $\vdash H \rightarrow I$, then $\vdash$I.
103. If $\vdash$H, then $\vdash \Box^d H$.
104. If $\vdash$H, then $\vdash \Diamond^i H$ where $e > i > 0$.
105. $\vdash (H \rightarrow^i I) \rightarrow (\Box^i H \rightarrow^i \Box^i I)$ where $i > 0$.
106. $\vdash (H \rightarrow^i I) \rightarrow (\Box^i H \rightarrow \Box^i I)$ where $e > i > 0$.
107. $\vdash \Box^j H \rightarrow \Box^i H$ where $(j - 1) = i > 0$.
108. $\vdash \Box^e H \rightarrow H$.
109. $\vdash \Box^i H \rightarrow \Box^i (x)H$ where $i > 0$.
110. $\vdash \Box^d H \rightarrow (R)H$.
111. $\vdash H \rightarrow (u)H$ where u has no free occurrence in H.
112. $\vdash (u)A \rightarrow H$ where H is like A except for having free occurrences of x or R, or occurrences of $\Box^i$ (with $e \geqslant i > 0$) wherever A has free occurrences of u.
113. $\vdash (u)(A \rightarrow B) \rightarrow ((u)A \rightarrow (u)B)$ where the fomula is a wff.

[1] L. Jonathan Cohen, *The Diversity of Meaning*, 2nd ed. p. 242.

114. If ⊢H, then ⊢(x)I where I is like H except for having free occurrences of x wherever H has free occurrences of y.

115. If ⊢H, then ⊢(R)I where I is like H except for having free occurrences of R wherever H has free occurrences of S.

116. If ⊢H, and $\square^{i}$ (with $e \geqslant i > 0$) occurs in H, and if, for some places in which $\square^{i}$ occurs in H and for all j such that $e \geqslant j > 0$, ⊢G where G is like H except for having occurrences of $\square^{j}$ in these places, then ⊢$(\square^{x})$A where A is like H except for having free occurrences of $\square^{x}$ in these places. (I.e. if '⊢. . .' is a metatheorem and '$\square^{i}$' occurs in one, or more, places in '. . .', with no restriction on i other than '$e \geqslant i > 0$', then '⊢$(\square^{x})$---' is a metatheorem, if '---' is like '. . .' except for having '$\square^{x}$' where, and only where, '$\square^{i}$' occurs in '. . .').

Metatheorems:– (Proofs will be omitted if they are of relatively unimportant theorems and sufficiently obvious).

117–133 describe some familiar derivative criteria of theoremhood, for wff, which will be referred to in later proofs. These criteria apply only where the formulas said to be provable are wffs.

117. If ⊢H & I, then ⊢H and ⊢I.

118. If ⊢H and ⊢I, then ⊢H & I.

119. If ⊢$H_1 \rightarrow H_2$, ⊢$H_2 \rightarrow H_3, \ldots,$⊢$H_{n-1} \rightarrow H_n$, then ⊢$H_1 \rightarrow H_n$.

120. ⊢$(u)H \rightarrow H$.

121. ⊢$H \rightarrow (\exists u)A$ where as in 112.

122. If ⊢H, then ⊢(u)H.

123. ⊢$(u)(A \,\&\, B) \leftrightarrow ((u)A \,\&\, (u)B)$.

124. ⊢$(u)(A \rightarrow B) \rightarrow ((\exists u)A \rightarrow (\exists u)B)$.

125. ⊢$(u)(A \rightarrow B) \rightarrow ((\exists u)A \rightarrow B)$ where u is not free in B.

126. ⊢$(u) -A \leftrightarrow -(\exists u)A$.

127. ⊢$-(u)A \leftrightarrow (\exists u)-A$.

128. ⊢$(u)(A \rightarrow B) \rightarrow ((\exists u)(A \,\&\, C) \rightarrow (\exists u)(B \,\&\, C))$.

129. ⊢$(u)(A \vee B) \rightarrow (A \vee (u)B)$ where u is not free in A.

130. ⊢$-(\exists u)A \rightarrow (u)(A \rightarrow B)$.

131. ⊢$(u)A \rightarrow (u)(B \rightarrow A)$.

132. ⊢$((u)(A \rightarrow B) \,\&\, (u)(B \rightarrow C)) \rightarrow (u)(A \rightarrow C)$.

133. ⊢$((u)(A \rightarrow B) \,\&\, (u)(A \rightarrow -B)) \rightarrow (u) -A$.

201–261 describe some fundamental logical properties of the primitive and defined modal operator-constants (squares).

201. ⊢$\square^{0}$ H — 101

202. ⊢$\square^{i}$ H $\rightarrow \square^{0}$ H — 101

203. If ⊢H, then ⊢$\square^{i}$ H.

204. ⊢$\square^{i}$ H $\rightarrow$ H where $i \geqslant e$.

205. If $\vdash H \rightarrow I$, then $\vdash \Box^i H \rightarrow \Box^i I$ and $\vdash \Diamond^i H \rightarrow \Diamond^i I$.
206. $\vdash (H \;\&\; (H \rightarrow^i I)) \rightarrow I$ where $i \geqslant e$.
207. $\vdash H \rightarrow \Diamond^i A$ where $i \geqslant e$.
208. $\vdash (H \rightarrow^j I) \rightarrow (\Box^i H \rightarrow \Box^i I)$ where $j \geqslant i$.
209. $\vdash H \rightarrow^i I) \rightarrow ((H \;\&\; J) \rightarrow^i I)$.
210. $\vdash (H \rightarrow^i I) \rightarrow (-I \rightarrow^i -H)$.
211. $\vdash ((H \rightarrow^i I) \;\&\; (I \rightarrow^i J)) \rightarrow (H \rightarrow^i J)$.
212. $\vdash ((H \rightarrow^i I) \;\&\; (H \rightarrow^i J)) \rightarrow ((H \rightarrow^i (I \;\&\; J))$.
213. $\vdash ((F \rightarrow^i G) \;\&\; (H \rightarrow^i I)) \rightarrow ((F \;\&\; H) \rightarrow^i (G \;\&\; I))$.
214. $\vdash \Box^i H \leftrightarrow (-H \rightarrow^i H)$.
215. $\vdash \Box^i H \rightarrow (I \rightarrow^i H)$.
216. $\vdash \Box^i -H \rightarrow (H \rightarrow^i I)$.
217. $\vdash \Box^i (H \;\&\; I) \rightarrow \Box^i H$ — 101, 203, 106, 102.
218. $\vdash (\Box^j H \;\&\; \Box^i I) \rightarrow \Box^i (H \;\&\; I)$ where $j \geqslant i$.
   i. $\vdash (\Box^j H \;\&\; \Box^i I) \rightarrow (\Box^i H \;\&\; \Box^i I)$ where $j \geqslant i$. — 107, 119, 101, 102.
   ii. $\vdash (\Box^i H \;\&\; \Box^i I) \rightarrow ((-(H \;\&\; I) \rightarrow^i H) \;\&\; (-(H \;\&\; I) \rightarrow^i I))$ — 215, 215, 118, 213, 102.
   iii. $\vdash (\Box^j H \;\&\; \Box^i I) \rightarrow (-(H \;\&\; I) \rightarrow^i (H \;\&\; I))$ where $j \geqslant i$ — i, ii, 212, 119.
   iv. $\vdash (\Box^j H \;\&\; \Box^i I) \rightarrow \Box^i (H \;\&\; I)$ where $j \geqslant i$ — iii, 214, 117, 119.
219. $\vdash (\Box^i H \;\&\; \Box^i I) \leftrightarrow \Box^i (H \;\&\; I)$.
220. $\vdash \Box^i H \rightarrow \Box^i (H \vee I)$.
221. $\vdash (\Box^j H \vee \Box^i I) \rightarrow \Box^i (H \vee I)$ where $j \geqslant i$.
222. $\vdash \Diamond^i H \rightarrow \Diamond^j H$ where $j \geqslant i$.
223. $\vdash$ If $\vdash -H$, then $\vdash -\Box^i H$ where $i > 0$.
   i. If $\vdash -H$, then $\vdash -\Box^i --H$ where $i \geqslant e$ — 207, 102.
   ii. $\vdash -\Box^i --H \rightarrow -\Box^i H$ — 101, 205, 101, 102.
   iii. If $\vdash -H$, then $\vdash -\Box^i H$ where $i > 0$ — i, 104, ii, 102.
224. $\vdash -\Box^i (H \;\&\; -H)$ where $i > 0$ — 101, 223.
225. $\vdash \Box^0 (H \;\&\; -H)$ — 201.
226. $\vdash -(\Box^i H \;\&\; \Box^j -H)$ where $i > 0 < j$ — 224, 218, 118, 101, 102.
227. $\vdash \Box^0 H \;\&\; \Box^0 -H$ — 101.
228. $\vdash \Box^i H \rightarrow \Diamond^j H$ where $i > 0 < j$ — 226, 101, 102.
229. $\vdash (H \rightarrow^j -I) \rightarrow (\Box^i H \rightarrow -\Box^j I)$ where $i > 0 < j$.
230. $\vdash (\Box^e H \vee \Box^e -H) \rightarrow (\Box^i H \rightarrow H)$ where $i > 0$
   i. $\vdash \Box^e H \rightarrow (\Box^i H \rightarrow H)$ — 108, 101, 119.
   ii. $\vdash \Box^e -H \rightarrow (\Box^i H \rightarrow H)$ where $i > 0$ — 226, 101, 102.
   iii. $\vdash (\Box^e H \vee \Box^e -H) \rightarrow (\Box^i H \rightarrow H)$ where $i > 0$ — i, ii, 118, 101, 102.

Thus, if the deterministic thesis that $\Box^e H \vee \Box^e -H$ were added to our criteria of theoremhood, we could infer with the aid of 230 that $\vdash \Box^i H \rightarrow H$ even where $e > i > 0$. So *this* deterministic thesis (which might conceivably be true even if we did not know it to be so) is incompatible with any concept of inductive support in terms of which we want to be able to say that there may sometimes be some inductive support even for a false hypothesis. In other words there is one clearly characterisable sense of 'determinism' in which inductive reasoning does not presuppose determinism, but is incompatible with it. Moreover, this assumption, that some propositions are such that neither they nor their negations are fully established (laws of nature), is precisely analogous to the assumption on which S4 was shown in §21 (p. 212) to be preferable to S5, viz. the assumption that some propositions are logically contingent. So there is corroboration here for mapping the logical syntax of inductive support on to a generalisation, or class of generalisations of S4 rather than of S5. It is obviously possible to go somewhat further in investigating the relation of different modal systems to inductive logic, since several systems are known to include S4 without including S5. But it seems doubtful whether any philosophical arguments can be given for further principles of inductive syntax than those that are derivable by generalising S4.

231. $\vdash \Box^{i}(H \,\&\, I) \rightarrow -(H \rightarrow^{i} - I)$ where $i > 0$

232. $\vdash (\Box^{j} H \rightarrow^{d} \Box^{i} H) \vee (\Box^{i} I \rightarrow^{d} \Box^{j} I)$

i. $\vdash -(\Box^{j} H \rightarrow^{d} \Box^{i} H) \rightarrow (\Box^{i} I \rightarrow^{d} \Box^{j} I)$ where $i > j > 0$ — 107, 119, 103, 101, 102.

ii. $\vdash -(\Box^{j} H \rightarrow^{d} \Box^{i} H) \rightarrow (\Box^{i} I \rightarrow^{d} \Box^{j} I)$ where $j > i > 0$ — 107, 119, 103, 101, 102.

iii. $\vdash -(\Box^{j} H \rightarrow^{d} \Box^{i} H) \rightarrow (\Box^{i} I \rightarrow^{d} \Box^{j} I)$ where $i = 0$ or $j = 0$ — 101, 103, 101, 102.

iv. $\vdash -(\Box^{j} H \rightarrow^{d} \Box^{i} H) \rightarrow (\Box^{i} I \rightarrow^{d} \Box^{j} I)$ where $i = j$ — 101, 103, 101, 102.

v. $\vdash (\Box^{j} H \rightarrow^{d} \Box^{i} H) \vee (\Box^{i} I \rightarrow^{d} \Box^{j} I)$ — i, ii, iii, iv.

233. $\vdash (\Diamond^{j} H \rightarrow^{d} \Diamond^{i} H) \vee (\Diamond^{i} I \rightarrow^{d} \Diamond^{j} I)$.

234. $\vdash \Box^{j} H \rightarrow \Box^{j} \Box^{i} H$ where $j \geqslant i$.

i. $\vdash ((I \vee -I) \rightarrow^{j} H) \rightarrow ((\Box^{j} (I \vee -I) \rightarrow^{j} \Box^{j} H) \,\&\, (\Box^{j} H \rightarrow^{j} \Box^{i} H))$ where $j \geqslant i$ — 107, 119, 203, 105, 118, 101, 102.

ii. $\vdash ((I \vee -I) \rightarrow^{j} H) \rightarrow (\Box^{j}(I \vee -I) \rightarrow^{j} \Box^{i} H)$ where $j \geqslant i$ — i, 211, 119.

iii. $\vdash \Box^{j} H \rightarrow (\Box^{j} \Box^{j}(I \vee -I) \rightarrow \Box^{j} \Box^{i} H)$ where $j \geqslant i$ — 215, ii, 208, 119.

iv. $\vdash \Box^{j} \Box^{j}(I \vee -I)$ — 101, 203, 203.

v. $\vdash \Box^{j} H \rightarrow \Box^{j} \Box^{i} H$ where $j \geqslant i$ — iii, iv, 118, 101, 102.

235. $\vdash \Box^{j} \Box^{i} H \rightarrow \Box^{j} H$ where $i \geqslant e$.

236. $\vdash \Diamond^{j} \Box^{i} H \rightarrow \Diamond^{j} H$ where $i \geqslant e$.

237. $\vdash \Box^{i} H \rightarrow (x)H$ where $i \geqslant e$ 109, 107, 108, 119.

238. $\vdash (x)\Box^{i} H \rightarrow \Box^{i}(x)H$.

i. $\vdash (x)\Box^{i} H \rightarrow (x)\Box^{i}(x)H$ 109, 122, 113, 102.

ii. $\vdash (x)\Box^{i} H \rightarrow \Box^{i}(x)H$ i, 120, 119.

239. $\vdash \Box^{i}(x)H \rightarrow (x)\Box^{i} H$.

i. $\vdash (x)(\Box^{i}(x)H \rightarrow \Box^{i} H)$ 120, 205, 114.

ii. $\vdash \Box^{i}(x)H \rightarrow (x)\Box^{i} H$ i, 113, 102, 111, 119.

240. $\vdash (\exists x)\Box^{i} H \rightarrow \Box^{i}(x)H$ 109, 122, 125, 102.

241. $\vdash \Diamond^{i} H \rightarrow (x)\Diamond^{i} H$

i. $\vdash (\exists x)\Box^{i} H \rightarrow \Box^{i} H$ 120, 205, 240, 119.

ii. $\vdash \Diamond^{i} H \rightarrow (x)\Diamond^{i} H$ i, 101, 102; 126, 117, 119.

242. $\vdash \Diamond^{i} (x)H \rightarrow (x)\Diamond^{i} H$ 120, 205, 241, 119.

243. $\vdash \Box^{i} H \leftrightarrow \Box^{i} I$ if I differs from H just in having y_1 free where H has x_1 (i.e. where H has x_1 free), y_2 free where H has $x_2, \ldots, y_n$ free where H has x_n. 109, 112, 205, 119, 118.

244. $\vdash (x_1)(x_2)\ldots(x_n)\Box^{i} H \leftrightarrow \Box^{i} I$ if as in 243.
112, 243, 117, 119; 243, 117, 109, 239, 119; 118.

245. $\vdash \Box^{d} H \leftrightarrow \Box^{d} I$ where I differs from H just in having S_1 free where H has R_1, S_2 free where H has $R_2, \ldots, S_n$ free where H has R_n.

246. $\vdash (R_1)(R_2)\ldots(R_n)\Box^{d} H \leftrightarrow \Box^{d} I$ if as in 245.

247. $\vdash \Box^{i}(R)H \rightarrow (R)\Box^{i} H$.

248. Let an occurrence of E be an *i*th grade occurrence in H, where $i \neq e$, if and only if it (i) forms a proper part of the occurrence in H of a part $\Box^{i}$ J of H, and (ii) does not form a proper part of the occurrence in H of a part $\Box^{j}$ G of H, where $j \geqslant i$, or of a part $\Box^{x}$ G. Let an occurrence of E be an *e*th grade occurrence in H if and only if it (i) forms a proper part of the occurrence in H of a part $\Box^{e}$ J or $\Box^{x}$ G, and (ii) does not form a proper part of the occurrence in H of a part $\Box^{j}$ G of H, where $j > e$. Then if E, F, I are such that I results from H by the substitution of F for one or more *i*th grade occurrences of E in H, and $u_1, u_2, \ldots u_n$ is a complete list of symbols occurring freely in E and F, then $\vdash (u_1)(u_2)\ldots(u_n)$ $(E \leftrightarrow^{j} F) \rightarrow (H \leftrightarrow^{i} I)$ where $j \geqslant i$. (The proof is analogous to that for the corresponding theorem[1] in quantified S4.) Thus inductive support operators generate a hierarchy of non-extensional substitutivity.

We are now in a position to explore the logical syntax of the monadic support-functor defined contextually in 9–16, developing first some metatheorems about zero grade support. (It should be remembered that $\Box^{x}$ cannot occur unbound in E, F, G, H, I or J).

[1] R. C. Barcan, op. cit., p. 13 ff.

301. $\vdash s[H] \geqslant 0$ 201.

302. $\vdash s[H] = 0 \leftrightarrow -(\exists\Box^x)\Box^x H$

i. $-(\exists\Box^x)\Box^x H \rightarrow s[H] = 0$ 121, 101, 102.

ii. $(\Box^x)-\Box^i H \rightarrow (\Box^x)-\Box^x H$ where $i = 1$
107, 101, 102; 116, 113, 102.

iii. $s[H] \leqslant 0 \rightarrow -(\exists\Box^x)\Box^x H$ 111, ii, 126, 119.

iv. $\vdash s[H] = 0 \leftrightarrow -(\exists\Box^x)\Box^x H$ i, iii, 118, 101, 102.

303. $\vdash s[H] > 0 \leftrightarrow (\exists\Box^x)\Box^x H$

i. $\vdash s[H] > 0 \rightarrow (\exists\Box^x)\Box^x H$ 101, 121, 119.

ii. $\vdash(\exists\Box^x)\Box^x H \rightarrow s[H] > 0$ 101, 107, 116, 125, 102; 101, 102.

iii. $\vdash s[H] > 0 \leftrightarrow (\exists\Box^x)\Box^x H$ i, ii, 118.

304. $\vdash s[H] = 0 \leftrightarrow s[H] \leqslant 0$

305. $\vdash s[H] = 0 \rightarrow s[I] \geqslant s[H]$ 302, 117, 130, 119.

306. $\vdash s[H] > 0 \rightarrow s[-H] = 0$

i. $\vdash(\Box^x)\Box^i H \rightarrow (\Box^x) -\Box^x -H$ where $i > 0$ 228, 116, 113, 102.

ii. $\vdash\Box^i H \rightarrow -(\exists\Box^x)\Box^x -H$ where $i > 0$ 111, i, 126, 117, 119.

iii. $\vdash(\exists\Box^x)\Box^x H \rightarrow -(\exists\Box^x)\Box^x -H$ ii, 116, 125, 102.

iv. $\vdash s[H] > 0 \rightarrow s[-H] = 0$ 303, 117, iii, 302, 117, 119.

307. $\vdash(H \rightarrow^d -I) \rightarrow (s[H] > 0 \rightarrow s[I] = 0)$

308. $\vdash s[H] = 0 \text{ v } s[-H] = 0$

i. $\vdash-\Box^i H \text{ v } (\Box^x) -\Box^x -H$ where $i > 0$
101, 228, 119; 116, 129, 102.

ii. $\vdash(\Box^x)-(\Box^x H) \text{ v } (\Box^x)-\Box^x -H$ i, 116, 129, 102.

iii. $\vdash s[H] = 0 \text{ v } s[-H] = 0$ ii, 126, 302, 118, 101, 102.

309. $\vdash(H \rightarrow^d -I) \rightarrow (s[H] = 0 \text{ v } s[I] = 0)$

310. $\vdash s[-H] \geqslant s[H] \rightarrow s[H] = 0$

311. $\vdash s[H] > 0 \rightarrow (s[H] \geqslant s[I] \rightarrow s[-I] \geqslant s[-H])$.

312. $\vdash(H \rightarrow^d I) \rightarrow (s[H] > 0 \rightarrow s[I] > 0)$ 208, 101, 102.

313. $\vdash(s[H] > 0 \;\&\; s[I] > 0) \rightarrow s[H \;\&\; I] > 0$ 218, 101, 102.

314. $\vdash(s[(x)(H \rightarrow I)] > 0 \;\&\; s[(x)(H \rightarrow -I)] > 0) \rightarrow s[(x)-H] > 0$
133, 103, 312, 102; 313, 119.

315. $\vdash s[H] > s[I] \rightarrow s[H] > 0$

316. $\vdash s[H] = 0 \rightarrow s[H \;\&\; I] = 0$.

317. $\vdash s[H \text{ v } I] = 0 \rightarrow s[H] = 0$.

318. $\vdash\Box^d H \rightarrow s[H] > 0$

319. $\vdash\Box^d -H \rightarrow s[H] = 0$.

We pass now to some general theorems about the monadic support-functor:

320. $\vdash s[H] \geqslant s[H]$

321. $\vdash s[H] \geqslant e \rightarrow H$

322. $\vdash s[H] \geqslant s[-H] \rightarrow s[I] \geqslant s[-H]$.

323. $\vdash s[H] \geqslant i \vee s[H] \leqslant i$ 107, 101, 102.

324. $\vdash s[H] \geqslant s[I] \vee s[I] \geqslant s[H]$

i. $\vdash(\Box^i H \& \Box^j H) \rightarrow ((\Box^i I \rightarrow \Box^i H) \vee (\Box^j H \rightarrow \Box^j I))$ 101.

ii. $\vdash(\Box^i H \& -\Box^j H) \rightarrow ((\Box^i I \rightarrow \Box^i H) \vee (\Box^j H \rightarrow \Box^j I))$ 101.

iii. $\vdash(-\Box^i H \& -\Box^j H) \rightarrow ((\Box^i I \rightarrow \Box^i H) \vee (\Box^j H \rightarrow \Box^j I))$ 101.

iv. $\vdash(-\Box^i H \& \Box^j H) \rightarrow -(\Box^i I \& -\Box^j I)$ 232, 204, 118; 101, 102.

v. $\vdash-(\Box^i H \& \Box^j H) \rightarrow ((\Box^i I \rightarrow \Box^i H) \vee (\Box^j H \rightarrow \Box^j I))$ iv, 101, 119.

vi. $\vdash(\Box^i I \rightarrow \Box^i H) \vee (\Box^j H \rightarrow \Box^j I)$ i, ii, iii, v, 118; 101, 102.

vii. $\vdash s[H] \geqslant s[I] \vee s[I] \geqslant s[H]$ vi, 116, 129, 102; 101, 102; 116, 129, 102.

325. $\vdash s[H] > i \rightarrow s[H] \geqslant i.$

326. $\vdash s[H] < i \rightarrow s[H] \leqslant i.$

327. $\vdash s[H] = i \rightarrow s[H] \geqslant i.$

328. $\vdash s[H] = i \rightarrow s[H] \leqslant i.$

329. $\vdash s[H] \geqslant j \rightarrow s[H] > i$ where $j > i$.

330 $\vdash s[H] > s[I] \rightarrow s[H] \geqslant s[I].$

Some general principles of transitivity are provable for comparisons between different grades of inductive support. For example, for the relation of 'greater than or equal to' we have:

331. $\vdash(s[H] \geqslant s[I] \& s[I] \geqslant s[J]) \rightarrow s[H] \geqslant s[J].$

332. $\vdash(s[H] \geqslant i \& s[I] \leqslant i) \rightarrow s[H] \geqslant s[I].$

333. $\vdash(s[H] \geqslant s[I] \& s[I] \geqslant i) \rightarrow s[H] \geqslant i.$

334. $\vdash(s[H] \leqslant i \& s[H] \geqslant s[I]) \rightarrow s[I] \leqslant i.$

For the relation of 'greater than' there are corresponding laws:

335. $\vdash(s[H] > s[I] \& s[I] > s[J]) \rightarrow s[H] > s[J].$

336. $\vdash(s[H] > i \& s[I] < i) \rightarrow s[H] > s[I].$

337. $\vdash(s[H] > s[I] \& s[I] > i) \rightarrow s[H] > i.$

338. $\vdash(s[H] < i \& s[H] > s[I] \rightarrow s[I] < i.$

For the relation of 'equal to':

339. $\vdash(s[H] = s[I] \& s[I] = s[J]) \rightarrow s[H] = s[J].$

340. $\vdash(s[H] = i \& s[I] = i) \rightarrow s[H] = s[I].$

341. $\vdash(s[H] = s[I] \& s[I] = i) \rightarrow s[H] = i.$

We also have such principles as:

342. $\vdash(s[H] > s[I] \& s[I] \geqslant s[J]) \rightarrow s[H] > s[J].$

343. $\vdash(s[H] \geqslant s[I] \& s[I] > s[J]) \rightarrow s[H] > s[J].$

The relationship between two hypotheses may impose certain restrictions on one hypothesis' grade of support relative to the other's:

344. $\vdash(\mathrm{H} \rightarrow^{\mathrm{i}} \mathrm{I}) \rightarrow \mathrm{s}[\mathrm{I}] \geqslant \mathrm{s}[\mathrm{H}]$ where $i \geqslant e$

i. $\vdash(\mathrm{H} \rightarrow^{\mathrm{i}} \mathrm{I}) \rightarrow (\Box^{\mathrm{j}}\, \mathrm{H} \rightarrow \Box^{\mathrm{j}}\, \mathrm{I})$ where $i \geqslant e \geqslant j$ 208

ii. $\vdash(\mathrm{H} \rightarrow^{\mathrm{i}} \mathrm{I}) \rightarrow \mathrm{s}[\mathrm{I}] \geqslant \mathrm{s}[\mathrm{H}]$ where $i \geqslant e$

i, 116, 113, 102; 111, 119.

This metatheorem embraces (for the monadic functor) what has been referred to in previous chapters as the consequence principle for hypotheses. Note that it makes no difference whether the principle's antecedent is underwritten by non-contingent assumptions or is established inductively.

345. $\vdash(\mathrm{H} \rightarrow^{\mathrm{i}} \mathrm{I}) \rightarrow \mathrm{s}[\mathrm{H}] > j \rightarrow \mathrm{s}[\mathrm{I}] > j$ where $i \geqslant e$.

346. $\vdash(\mathrm{H} \rightarrow^{\mathrm{i}} \mathrm{I}) \rightarrow (\mathrm{s}[\mathrm{H}] \geqslant j \rightarrow \mathrm{s}[\mathrm{I}] \geqslant j)$, where $i \geqslant e$.

As corollaries of 344 we mention 347–350.

347. $\vdash(\mathrm{H} \leftrightarrow^{\mathrm{i}} \mathrm{I}) \rightarrow \mathrm{s}[\mathrm{H}] = \mathrm{s}[\mathrm{I}]$ where $i \geqslant e$.

348. $\vdash \mathrm{s}[\mathrm{H}] \geqslant \mathrm{s}[\mathrm{H}\ \&\ \mathrm{I}]$.

349. $\vdash \mathrm{s}[\mathrm{H} \vee \mathrm{I}] \geqslant \mathrm{s}[\mathrm{H}]$.

350. $\vdash \mathrm{s}[\mathrm{H} \vee \mathrm{I}] \geqslant \mathrm{s}[\mathrm{H}\ \&\ \mathrm{I}]$.

351. $\vdash \mathrm{s}[\mathrm{H}] \geqslant \mathrm{s}[\mathrm{I}] \rightarrow \mathrm{s}[\mathrm{H}\ \&\ \mathrm{I}] = \mathrm{s}[\mathrm{I}]$.

This represents what has been referred to in previous chapters as the general conjunction principle (for the monadic functor).

i. $\vdash(\Box^{\mathrm{i}}\, \mathrm{I} \rightarrow \Box^{\mathrm{i}}\, \mathrm{H}) \rightarrow ((\Box^{\mathrm{i}}\, \mathrm{I} \rightarrow \Box^{\mathrm{i}}(\mathrm{H}\ \&\ \mathrm{I}))\ \&\ (\Box^{\mathrm{i}}(\mathrm{H}\ \&\ \mathrm{I}) \rightarrow \Box^{\mathrm{i}}\, \mathrm{I})))$

217, 218, 118, 101, 102.

ii. $\vdash \mathrm{s}[\mathrm{H}] \geqslant \mathrm{s}[\mathrm{I}] \rightarrow \mathrm{s}[\mathrm{H}\ \&\ \mathrm{I}] = \mathrm{s}[\mathrm{I}]$

i, 116, 113, 102; 123, 119.

352. $\vdash \mathrm{s}[\mathrm{H}\ \&\ \mathrm{I}] \geqslant \mathrm{s}[\mathrm{I}] \rightarrow \mathrm{s}[\mathrm{H}] \geqslant \mathrm{s}[\mathrm{I}]$ 217, 101, 102; 116, 113, 102.

353. $\vdash \mathrm{s}[\mathrm{H} \rightarrow \mathrm{I}] \geqslant \mathrm{s}[\mathrm{H}] \rightarrow \mathrm{s}[\mathrm{I}] \geqslant \mathrm{s}[\mathrm{H}]$ 208, 101, 102; 116, 113, 102.

354. $\vdash \mathrm{s}[\mathrm{H}] \geqslant e \rightarrow \mathrm{s}[\mathrm{H}] \geqslant \mathrm{s}[\mathrm{I}]$ 215, 208, 119; 116, 113, 102.

355. $\vdash \Box^{\mathrm{d}}\, \mathrm{H} \rightarrow \mathrm{s}[\mathrm{H}] \geqslant e$.

356. $\vdash \mathrm{s}([(x)(\mathrm{H} \rightarrow \mathrm{I}] \geqslant \mathrm{j}\ \&\ \mathrm{s}[(\mathrm{x})(\mathrm{H} \rightarrow -\mathrm{I})] \geqslant \mathrm{i}) \rightarrow \mathrm{s}[(x)\text{–}\mathrm{H}] \geqslant \mathrm{i}$

133, 205, 218, 119.

357. $\vdash \mathrm{s}[(x_1)(x_2)\ldots(x_{\mathrm{n}})\mathrm{H}] = \mathrm{s}[\mathrm{I}]$ if I differs from H just in having y_1 free where H has x_1, y_2 free where H has x_2, . . ., y_{n} free where H has x_{n}. This metatheorem represents, for the monadic functor, what has been referred to in previous chapters as the uniformity principle.

i. $\vdash \mathrm{s}[\mathrm{I}] \geqslant \mathrm{s}[(x_1)(x_2)\ldots(x_{\mathrm{n}})\mathrm{H}]$ where as above

112, 119, 103, 208, 102; 116, 113, 102.

ii. $\vdash s[(x_1)(x_2)\ldots(x_n)H] \geqslant s[I]$ where as above
243, 117, 109, 119; 116, 113, 102.

iii. $\vdash s[(x_1)(x_2)\ldots(x_n)H] = s[I]$ where as above i, ii, 118.

But we cannot prove $s[(R)H] = s[I]$ if I differs from H just in having S free where H has R: cf. §9, p. 82 f.

358. $\vdash s[(x_1)(x_2)\ldots(x_n)H] > s[(x_1)(x_2)\ldots(x_n)G] \leftrightarrow s[I] > s[J]$ if I differs from H as in 357, and J differs from G analogously in having z_1 free where G has x_1, etc. This metatheorem represents, for the monadic support-functor, what has been referred to in previous chapters as the instantial comparability principle.

i. $\vdash s[(x_1)(x_2)\ldots(x_n)H] > s[(x_1)(x_2)\ldots(x_n)G] \rightarrow s[I] > s[(x_1)(x_2)\ldots(x_n)G]$ if as above
357, 343, 118, 101, 102.

ii. $\vdash s[I] > s[(x_1)(x_2)\ldots(x_n)G] \rightarrow s[I] > s[J]$ where as above
357, 342, 118, 101, 102.

iii. $\vdash s[(x_1)(x_2)\ldots(x_n)H] > s[(x_1)(x_2)\ldots(x_n)G] \rightarrow s[I] > s[J]$ if as above i, ii, 119.

iv. $\vdash s[(x_1)(x_2)\ldots(x_n)H] > s[(x_1)(x_2)\ldots(x_n)G] \leftrightarrow s[I] > s[J]$ if as above iii, sim. iii, 118.

359. $\vdash s[H] = s[I]$ if as in 357. I.e. any two substitution-instances of the same first-order generalisation have the same grade of support as one another.

360. $\vdash s[I_1 \mathbin{\&} I_2 \mathbin{\&} \ldots \mathbin{\&} I_n] = s[I_1]$ if I_i differs from I_j just as I from H in 357. The proof of this instantial conjunction principle uses repeated applications of 359, 117 and 351.

We pass now to the syntax of the dyadic support-functor, defined in 17–24, and develop first some metatheorems about 0th grade support. The proofs are for the most part closely analogous to those of the corresponding metatheorems for the monadic support-functor, 301–319.

401. $\vdash s[H, E] \geqslant 0$. With 401 cf. Carnap's T 59–1a, viz. $0 \leqslant c(h,e) \leqslant 1$[1]. But $s[H,E] \leqslant e$ is not provable in the present system, because the non-existence of a higher grade of inductive support than e/e, where e is some definite positive integer, is conceived to be an empirical fact, depending on the number of inductively relevant variables (cf. §§7–8).

402. $\vdash s[H, E] = 0 \leftrightarrow -(\exists\Box^x)(E \rightarrow^e \Box^x H)$.

403. $\vdash s[H, E] > 0 \leftrightarrow (\exists\Box^x)(E \rightarrow^e \Box^x H)$.

404. $\vdash s[H, E] = 0 \leftrightarrow -s[H, E] \leqslant 0$.

405. $\vdash s[H, E] = 0 \rightarrow s[I, F] \geqslant s[H, E]$.

[1] *Logical Foundations of Probability*, p. 316.

406. $\vdash \Diamond^e E \rightarrow (s[H, E] > 0 \rightarrow s[-H, E] = 0)$.
   i. $\vdash (E \rightarrow^e (\Box^i H \& \Box^i -H)) \rightarrow \Box^e -E$ 226, 101, 102; 205.
   ii. $\vdash \Diamond^e E \rightarrow ((E \rightarrow^e \Box^i H) \rightarrow -(E \rightarrow^e \Box^i -H))$ 212, i, 118, 101, 102.
   iii. $\vdash \Diamond^e E \rightarrow (s[H, E] > 0 \rightarrow s[-H, E] = 0)$ ii, 101, 102.

   This principle may be contrasted with Carnap's T 59–1p, viz. $c(-h, e) = 1 - c(h, e)$.

407. $\vdash (\Diamond^e E \& (H \rightarrow^d -I)) \rightarrow (s[H, E] > 0 \rightarrow s[I, E] = 0)$
408. $\vdash \Diamond^e E \rightarrow (s[H, E] = 0 \text{ v } s[-H, E] = 0)$.
409. $\vdash (\Diamond^e E \& (H \rightarrow^d -I)) \rightarrow (s[H, E] = 0 \text{ v } s[I, E] = 0)$.
410. $\vdash \Diamond^e E \rightarrow (s[-H, E] \geqslant s[H, E] \rightarrow s[H, E] = 0)$. Contrast the principle that follows from Carnap's T 59–1p, viz. if $c(h,e) = c(-h,e)$ then $c(h, e) = \frac{1}{2}$.
411. $\vdash \Diamond^e E \rightarrow (s[H, E] > 0 \rightarrow (s[H, E] \geqslant s[I, E] \rightarrow s[-I, E] \geqslant s[-H, E])$.
412. $\vdash (H \rightarrow^d I) \rightarrow (s[H, E] > 0 \rightarrow s[I, E] > 0)$.
413. $\vdash (s[H, E] > 0 \& s[I, E] > 0) \rightarrow s[H \& I, E] > 0$.
414. $\vdash (s[(x)(H \rightarrow I), E] > 0 \& s[(x)(H \rightarrow -I), E] > 0) \rightarrow s[(x)-H, E] > 0$.
415. $\vdash (s[(x)(H \rightarrow I, E] > 0 \& s[(x)(H \rightarrow -I), E] > 0) \rightarrow (s[(\exists x)H, E] > 0) \rightarrow \Box^e -E)$

   I.e. if E gives positive support to genuinely inconsistent propositions, E cannot be true: cf. §8, p. 64 f., above. The proof uses 414 and 406.

416. $\vdash s[H, E] > s[I, E] \rightarrow s[H, E] > 0$.
417. $\vdash s[H, E] = 0 \rightarrow s[H \& I, E] = 0$. Cf. Carnap's T 59–6a, viz. if $c(h, e) = 0$, then $c(h \& i, e) = 0$.
418. $\vdash s[H \text{ v } I, E] = 0 \rightarrow s[H, E] = 0$. Cf. Carnap's T 59–6s, viz. if $c(h \text{ v } i, e) = 0$, then $c(h, e) = 0$.
419. $\vdash (s[H, E] > 0 \& s[-H, F] > 0) \rightarrow \Box^e -(E \& F)$.

We pass now to some general metatheorems about the dyadic support-functor. Where corresponding metatheorems exist for the monadic support-functor, the proofs are closely analogous.

420. $\vdash s[H, E] \geqslant s[H, E]$.
421. $\vdash (s[H, E] \geqslant e \& E) \rightarrow H$.
422. $\vdash s[H, \Box^j H] \geqslant$ i where $j \geqslant i$.
423. $\vdash s[H, E] \geqslant s[-H, E] \rightarrow s[I, E] \geqslant s[-H, E]$.
424. $\vdash s[H, E] \geqslant \text{i v } s[H, E] \leqslant \text{i}$.
425. $\vdash s[H, E] \geqslant s[I, F] \text{ v } s[I, F] \geqslant s[H, E]$.

Metatheorems 426–444 are like metatheorems 325–343 respectively, except for having 's[H,E]' in place of 's[H]', 's[I, F]' in place of 's[I]', and 's[J,G]' in place of 's[J]'. Metatheorems 445–455 are like metatheorems 344–354, respectively, except for having 's[. . .,E], in place of 's[. . .]' for each filler of the blank. With the consequence principle for hypotheses 445 (which is like 344) cf. Carnap's T 59–2d, viz. if $\vdash h_1 \rightarrow h_2$, or $\vdash e.h_1 \rightarrow h_2$, then $c(h_1,e) \leqslant c(h_2,e)$. With 447 (which is like 346) cf. Carnap's T 59–li, viz. if $\vdash h_1 \leftrightarrow h_2$, then $c(h_1,e) = c(h_2,e)$. With 448 (which is like 347) cf. Carnap's T 59–2e, viz. $c(h \; \& \; i,e) \leqslant c(h,e)$. With 449 (which is like 348) cf. Carnap's T 59–2f, viz. $c(h,e) \leqslant c(h \vee i,e)$. But with 452 (which is like 351) contrast Carnap's T 59–1n, viz. $c(h \; \& \; i,e) = c(h,e) \times c(i,e \; \& \; h)$.

In addition we have:

456. $\vdash s[H \rightarrow I, E] \geqslant e \rightarrow s[I, E] \geqslant s[H, E]$. Cf. Carnap's T 59–6f, viz. if $c(h_1 \rightarrow h_2, e) = 1$ then $c(h_1, e) \leqslant c(h_2, e)$.

457. $\vdash s[H \leftrightarrow I, E] \geqslant e \rightarrow s[I, E] = s[H, E]$. Cf. Carnap's T 59–6g, viz. if $c(h_1 \leftrightarrow h_2,e) = 1$ then $c(h_1, e) = c(h_2, e)$.

458. $\vdash (E \rightarrow^{i} F) \rightarrow s[H,E] \geqslant s[H,F]$ where $i \geqslant e$. This embraces the principle that has been referred to previously as the consequence principle for evidential propositions. Note that here too it makes no difference whether the principle's antecedent is underwritten by non-contingent assumptions or is established inductively.

459. $\vdash (E \leftrightarrow^{i} F) \rightarrow s[H, E] = s[H, F]$ where $i \geqslant e$. Cf. Carnap's T59–1h, viz. if $\vdash e_1 \leftrightarrow e_2$ then $c(h_1 e_1) = c(h_1, e_2)$.

460. $\vdash ((E \leftrightarrow^{i} F) \; \& \; (H \leftrightarrow^{j} I)) \rightarrow s[H, E] = s[I, F]$ where $i \geqslant e \leqslant j$. This embraces the principle that has been referred to previously as the equivalence principle (for the dyadic functor).

461. $\vdash s[H, E \; \& \; F] \geqslant s[H, E]$. On the intended interpretation there is no paradox here: cf. §8, p. 62, above.

462. $\vdash (s[H, E] \geqslant i \; \& \; s[H, F] \geqslant i) \rightarrow s[H, E \; \& \; F] \geqslant i$. Though mere multiplicity of instances—repetition of precisely the same test-results—does not serve to increase inductive support, we cannot prove $(s[H,E] = i \; \& \; s[H,F] = i) \rightarrow s[H,E \; \& \; F] = i$, because E might report one part of a successful result from test t_j, where $j > i$, and F might report the remainder of this result. Compare, however, 499 below.

463. $\vdash s[H, E] = s[I, F] \rightarrow s[H \; \& \; I, E \; \& \; F] \geqslant s[H, E]$.

464. $\vdash (\Diamond^{e} E \; \& \; (E \rightarrow^{i} -H)) \rightarrow s([H, E] < e$ where $i \geqslant e$.

i. $((E \rightarrow^{i} -H) \; \& \; (E \rightarrow^{i} H)) \rightarrow \Box^{e} -E$ where $i \geqslant e$
101, 205, 107, 218, 119.

ii. $\vdash -(E \rightarrow^{i} H) \rightarrow -(E \rightarrow^{i} \Box^{e} H)$ where $i \geqslant e$
204, 211, 118, 101, 102.

iii. $\vdash (\Diamond^{e} E \; \& \; (E \rightarrow^{i} -H)) \rightarrow s[H, E] < e$ where $i \geqslant e$
i, ii, 118, 101, 102.

Thus if E reports a genuine counter-instance to H, H cannot attain the status of a law on the evidence of E, since, according to E, it does not survive the manipulation of all relevant variables. Contrast Carnap's rather stronger theorem T 59–1e, viz. if $\vdash e \rightarrow -h$ then $c(h, e) = 0$. The principle provable here is weaker in order to allow for the fact that even if the test-results reported in E include a genuine counter-instance to H, where H is universal, E may still not exclude H from being supported by the results of some less thorough test. Even where H is singular, and contradicted by E, E may still not exclude there being some inductive support for the generalisation instantiated by H and thus derivatively for H itself. However, the yet weaker thesis, that H can sometimes attain the status of a law on the evidence of E, even when E includes the report of a genuine counter-instance to H, is to be rejected. For logical purposes we must suppose that what attains the status of a law in such cases is either a less simple version of H or some proposition that is deducible from such a version: cf. §16 above.

465. $\vdash \Diamond^e \Box^e E \rightarrow ((E \rightarrow^i -H) \rightarrow s[H, E] = 0)$ where $i \geqslant e$

i. $\vdash (\Box^e E \rightarrow (\Box^j H \,\&\, \Box^e -H)) \rightarrow -\Box^e E$ 226, 101, 102.

ii. $\vdash \Diamond^e \Box^e E \rightarrow -(\Box^e E \rightarrow^e (\Box^j H \,\&\, \Box^e -H))$ i, 205, 101, 102.

iii. $\vdash \Diamond^e \Box^e E \rightarrow ((\Box^e E \rightarrow^e \Box^e -H) \rightarrow -(\Box^e E \rightarrow^e \Box^j H))$ ii, 212, 118, 101, 102.

iv. $\vdash \Diamond^e \Box^e E \rightarrow ((E \rightarrow^i -H) \rightarrow -(\Box^e E \rightarrow^e \Box^j H))$ where $i \geqslant e$ 107, 105, 119; iii, 118; 101, 102.

v. $\vdash (E \rightarrow^e \Box^j H) \rightarrow (\Box^e E \rightarrow^e \Box^j H)$ 108, 101, 102, 205.

vi. $\vdash \Diamond^e \Box^e E \rightarrow ((E \rightarrow^i -H) \rightarrow -(E \rightarrow^e \Box^j H))$ where $i \geqslant e$ iv, v, 118, 101, 102.

vii. $\vdash \Diamond^e \Box^e E \rightarrow ((E \rightarrow^i -H) \rightarrow s[H, E] = 0)$ where $i \geqslant e$ vi, 101, 102.

I.e., if there is no law barring E itself from being a law, then a hypothesis can derive no support at all from E if its falsehood is implied by E. But a canonical test-report has the form $(\exists x)(Rx \,\&\, Sx \,\&\, T_i x \,\&\, \ldots)$ and so can neither be a testable hypothesis itself nor logically deducible from one (see §7 and §11 above). Hence the intended interpretation of the calculus implies that $\Diamond^e \Box^e E$ is normally false where E is an evidential report. $\Diamond^e \Box^e E$ may be true where E is equivalent to a universally quantified conditional, or a substitution-instance of one. But then H would not be a proposition of the kind that can enjoy inductive support.

Similarly, there is no analogue for Carnap's T 59–1b, viz. if $\vdash e \rightarrow h$ then $c(h, e) = 1$, since the domain of the dyadic inductive support-functor is not the interval between E's contradicting H and E's implying H, but rather the interval between E's giving no support at all to H and E's implying H to be a law or a consequence of one. Metatheorems 466–468 are the nearest we get to Carnap's T 59–1b.

466. $\vdash(E \rightarrow^{i} H) \rightarrow (\Box^{i} E \rightarrow s[H, E] \geqslant e)$ where $i \geqslant e$.

467. $\vdash(E \rightarrow^{i} H) \rightarrow s[H, \Box^{i} E] \geqslant e$ where $i \geqslant e$.

468. $\vdash s[H, E] \geqslant e \rightarrow (E \rightarrow^{e} H)$.

469. $\vdash\Box^{d} H \rightarrow s[H, E] \geqslant e$. Cf. Carnap's T 59–1c, viz. if h is L-true then c(h, e) = 1.

470. $\vdash\Box^{i} -H \rightarrow (\Diamond^{e} E \rightarrow s[H, E] = 0)$ where $i \geqslant e$. Cf. Carnap's T 59–1f, viz. if h is L-false then c(h, e) = 0.

471. $\vdash\Box^{i} E \rightarrow s[H, E \ \& \ F] = s[H, F]$ where $i \geqslant e$. Provable with the aid of 215. Cf. Carnap's notion[1] of null confirmation on tautological evidence.

472. $\vdash\Box^{i} -E \rightarrow s[H, E] \geqslant e$ where $i \geqslant e$. Cf. Carnap's argument[2] that if any value at all is to be assigned to c(h, e) where e is L-false the assignment should be c(h, e) = 1.

473. $\vdash(s[H, E] > 0 \ \& \ s[H, E] < i) \rightarrow s[-H, E] = 0$. This is a corollary of 477, in virtue of 406.

474. $\vdash s[H, E] > 0 \rightarrow (\Diamond^{e} E \rightarrow \Diamond^{e} H)$. The proof uses 228 and 234.

475. $\vdash s[(x_1)(x_2)\ldots(x_n)H, E] = s[I, E]$ if as in 357.

476. $\vdash s[(x_1)(x_2)\ldots(x_n)H, E] > s[(x_1)(x_2)\ldots(x_n)G, F] \rightarrow s[I, E] > s[J, F]$ if as in 358.

This formal counterpart of the instantial comparability principle (for the dyadic functor) makes it clear that the principle does not need the restriction on evidential propositions that was postulated in §2. Since logical implication does not entail inductive support, and all inductive support is mediated through generalisations, it is even possible for J and F to be identical here.

477. $\vdash s[I_1 \ \& \ I_2 \ \& \ \ldots \ \& \ I_n, E] = s[I_1, E]$ if as in 360.

478. $\vdash s[I, E] = s[H, E]$ if as in 357. The proof uses 477.

479. $\vdash s[G, \Box^{i} I] = s[G, \Box^{i} H]$ if as in 357. The proof uses 243.

480. $\vdash s[(x_1)(x_2)\ldots(x_n)H, E] = s[(x_1)(x_2)\ldots(x_n)H, E \ \& \ F) \leftrightarrow s[I, E] = s[I, E \ \& \ F]$ if as in 357. I.e., evidence does not increase the grade

[1] R. Carnap, op. cit., p. 289, 307 ff.

[2] Ibid. p. 295. Philosophers who do not like such principles as those expressed in metatheorems 215, 216, 471 and 472 could avoid them by building their inductive logic on a generalisation of the so-called 'calculus of entailment', constructed by A. R. Anderson and N. D. Belnap, which is a proper sub-system of S4: see G. E. Hughes and M. J. Cresswell, *An Introduction to Modal Logic* (1968), p. 299 ff. But on this subject cf. L. Jonathan Cohen, op. cit., p. 272 ff.

of support for a universal hypothesis if and only if it does not increase the grade of support for any substitution-instance of that hypothesis.

481. $\vdash$s[H, E & F] > s[H, E] $\rightarrow$ s[I, E & F] > [I, E] if as in 357. I.e., a proposition increases the grade of support for one substitution-instance of a hypothesis only if it also increases the grade of support for any other.

We may turn now to the relations between the monadic support-functor and the dyadic one.

482. $\vdash$s[H] $\geqslant e \rightarrow$ s[H, E] $\geqslant e$. This metatheorem, provable with the aid of 215, is the inductive analogue of one of the so-called paradoxes of strict implication, viz. the thesis that a logical truth is logically implied by any proposition. Inductive researches are neither hampered nor assisted by this type of paradox since it just entitles us to infer a certain type of conclusion (from any evidence) if we already know its truth.

483. $\vdash$s[–H] $\geqslant e \rightarrow (\Diamond^{e}$ E $\rightarrow$ s[H, E] $= 0)$. This is a corollary of 482, in virtue of 406.

484. $\vdash$s[H, E] $> 0 \rightarrow$ (s[–H] $\geqslant e \rightarrow$ s[–E] $\geqslant e$).

485. $\vdash$s[E] $\geqslant e \rightarrow$ s[H, E & F] = s[H, F]. This metatheorem, also provable with the aid of 215, excludes established laws from being evidence that increases the support for any hypothesis, since a statement of supporting evidence is normally a conjunction of existential propositions that reports one or more canonical test-results, as in §7. This is a restatement of 471.

486. $\vdash$s[–E] $\geqslant e \rightarrow$ s[H, E] $\geqslant e$. This is a restatement of 472.

487. $\vdash$(s[H, E] > 0 & s[H] $= 0) \rightarrow$ –E.

488. $\vdash$s[H, E] $\geqslant i$ & s[I, H] $\geqslant i) \rightarrow$ s[s[I] $\geqslant i$, E] $\geqslant i$. This represents the nearest approach, as it were, to transitivity that inductive inference achieves, where $e > i$.

Under certain conditions inductive detachment is possible, as in 489–490:

489. $\vdash$(s[H, E] $\geqslant i$ & E) $\rightarrow$ s[H] $\geqslant i$.
490. $\vdash$(s[H, E] $> i$ & E) $\rightarrow$ s[H] $> i$.

Detachment is not possible where we have s[H, E] $\leqslant i$, s[H, E] $< i$, or s[H, E] $= i$, since we cannot infer from the failure of one true proposition E to give more than such-or-such a grade of support to H that no other proposition giving greater support is true. But certain inferences in the other direction are possible, as in 491–499:

491. $\vdash(E \rightarrow^{e} s[H] \geqslant i) \rightarrow s[H, E] \geqslant i$.
492. $\vdash(E \rightarrow^{e} s[H] > i) \rightarrow s[H, E] > i$.
493. $\vdash(E \rightarrow^{e} s[H] \leqslant i) \rightarrow (\Diamond^{e} E \rightarrow s[H, E] \leqslant i)$.
494. $\vdash(E \rightarrow^{e} s[H] < i) \rightarrow (\Diamond^{e} E \rightarrow s[H, E] < i)$.
495. $\vdash(E \rightarrow^{e} s[H] = i) \rightarrow (\Diamond^{e} E \rightarrow s[H, E] = i)$.
496. $\vdash(E \rightarrow^{e} s[H] \geqslant s[I]) \rightarrow s[H, E] \geqslant s[I, E]$.
497. $\vdash(E \rightarrow^{e} s[H] = s[I]) \rightarrow s[H, E] = s[I, E]$.
498. $\vdash s[H, E] \geqslant i \leftrightarrow s[E \rightarrow s[H] \geqslant i] \geqslant e$.
499. $\vdash((E \rightarrow^{e} s[H] = i) \ \&\ (F \rightarrow^{e} s[H] = i)) \rightarrow (\Diamond^{e}(E \ \&\ F) \rightarrow s[H, E \ \&\ F] = i)$. Thus mere multiplicity of instances—repetition of precisely the same test-result—does not serve to increase inductive support.

Finally, some properties of the inductive information-functor i[,], contextually defined in 25–32 above are developed. Several of these are analogous to properties of the dyadic support-functor, e.g.:

501. $\vdash i[E, H] \geqslant 0$.
502. $\vdash i[E, H] \geqslant i[E, H]$.
503. $\vdash i[E, H] \geqslant i \vee i[E, H] \leqslant i$.
504. $\vdash i[E, H] \geqslant i[F, I] \vee i[F, I] \geqslant i[E, H]$.
505. $\vdash i[E, H] = 0 \rightarrow i[F, I] \geqslant i[E, H]$.
506. $\vdash(i[E, H] \geqslant i[F, I] \ \&\ i[F, I] \geqslant i[G, J]) \rightarrow i[E, H] \geqslant i[G, J]$.
507. $\vdash(i[E, H] = i \ \&\ i[F, I] = i) \rightarrow i[E, H] = i[F, I]$.
508. $\vdash(H \leftrightarrow^{i} I) \rightarrow i[E, H] = i[E, I]$ where $i \geqslant e$. But there is no analogue for 445.
509. $\vdash(E \rightarrow^{i} F) \rightarrow i[E, H] \geqslant i[F, H]$ where $i \geqslant e$.
510. $\vdash(E \leftrightarrow^{i} F) \rightarrow i[E, H] = i[F, H]$ where $i \geqslant e$.
511. $\vdash(i[E, H] = i \ \&\ i[E, I] = i) \rightarrow i[E, H \ \&\ I] = i$.
512. $\vdash i[E, H] \geqslant i[E, I] \rightarrow i[E, H \ \&\ I] \geqslant i[E, I]$.
513. $\vdash i[E \ \&\ F, H] \geqslant i[E, H]$.
514. $\vdash s[H] \geqslant e \rightarrow i[E, H] \geqslant e$.
515. $\vdash s[E] \geqslant e \rightarrow i[E \ \&\ F, H] = i[F, H]$.
516. $\vdash s[-E] \geqslant e \rightarrow i[E, H] \geqslant e$.

But the inductive information-functor also has some more idiosyncratic properties, e.g.:

517. $\vdash i[E, H] = i[E, -H]$.
518. $\vdash i[E, H] \geqslant e \rightarrow ((E \rightarrow^{e} H) \vee (E \rightarrow^{e} -H))$.
519. $\vdash s[-H] \geqslant e \rightarrow i[H, E] \geqslant e$.
520. $\vdash i[E, H] < e \rightarrow (\Diamond^{e} H \ \&\ \Diamond^{e} -H)$.
521. $\vdash i[E, H] = 0 \rightarrow i[H, E] = i[-H, E]$.
522. $\vdash s[H, E] = s[-H, E] \rightarrow (\Diamond^{e} E \rightarrow i[E, H] = 0)$.
523. $\vdash s[H, E] = i \rightarrow i[E, H] = i$ where $i > 0$.

In §7 it was argued that inductive support-functions should be conceived to take finite ordinals as values. Hence, if an inductive calculus is to admit of its intended interpretation, the contextually defined functors of the formal calculus must satisfy standard requirements for expressions denoting the members of a well-ordered set.

The standard requirements for membership of a well-ordered set S may be listed as follows:

i. For any member x, $x \geqslant x$.
ii. For any members x and y, if $x \geqslant y$ and $y \geqslant x$, then $x = y$.
iii. For any members x, y and z, if $x \geqslant y$ and $y \geqslant z$, then $x = z$.
iv. For any members x and y, either $x \geqslant y$ or $y \geqslant x$.
v. There is a member y, such that for any member x, $x \geqslant y$.
vi. For any subset S' of S, there is a member y of S', such that for any member x of S', $x \geqslant y$.

In the case of the monadic (dyadic) support-functor requirement i is given by 320 (420); requirement ii by 13 and 16 (21 and 24); requirement iii by 331–334 (432–435); requirement iv by 323 and 324 (424 and 425); and requirement v by 301 (401). Satisfaction of requirement vi is ensured by 301 (401), where 0 is a member of S'; and, where 0 is not a member of S', it is ensured by the fact that the superscripts of the primitive modal operator-constants represented by the square in the definiens of 9 (17) and in that of 11 (19), and quantified over in the expressions represented by the definiens of 14 (22), denote only positive integers. These requirements are also demonstrably satisfied by the inductive information-functor.

It is demonstrable that any inductive calculus is consistent if standard second-order quantification-theory is consistent.[1] To interpret wff of such a calculus as wff of standard quantification-theory, just replace each occurrence of $\Box^{i}A$ or $\Box^{x}A$ in a wff by $(R_1)(R_2)\ldots(R_n)(x_1)(x_2)\ldots(x_n)A$, where $R_1\ R_2, \ldots, R_n, x_1, x_2, \ldots, x_n$ are all the predicate and individual symbols free in A, and each occurrence of $(\exists\Box^{x})A$ by $(\exists y)A$ where y is not free in A. As criteria of theoremhood 101, 102, 111, 112, 113, 114 and 115, where u is x or R, will now suffice since 103–110, 116, and also 111, 112 and 113 where u is $\Box^{x}$, are now demonstrable from the other criteria. Specifically, 103 asserts that if $\vdash$A then $\vdash(R_1)\ldots(x_n)A$ where as above, and this is provable from 114 and 115; and 104 now asserts that if $\vdash$A then $\vdash -(R_1)\ldots(x_n)-A$, which is provable with the help of 112. Similarly 105–107 are easily proved with the help of 113 and 114; 108 with that of 112; and 109 with that of 111, 113 and 114. To prove 116, and

[1] On the consistency of second-order quantification-theory cf. A. Church, *Introduction to Mathematical Logic* (1956), vol. I, p. 306 f.

also 111, 112 and 113 where u is $\Box^{x}$, we first prove $H \rightarrow (\exists x)H$, where x is not free in H, from 111 and 112. But if, with the specified replacements, 101–102 and 111–115 (where u is x or R) suffice as criteria of theoremhood, we need no other criteria now than those of standard second-order quantification-theory. Any calculus of the kind proposed, for the logical syntax of monadic and dyadic support-functors, is mappable thus on to standard second-order quantification-theory and must be consistent if the latter is.

This mapping does not itself exclude the possibility that a contradiction may be generated by the semantical rules which are used to read off the principles of inductive syntax from the theorems of such a calculus. But all familiar semantical antinomies have been excluded in the usual way by a language-level requirement; and familiar modal paradoxes have been excluded by the requirement of a one-one correspondence between each kind of logico-linguistic role in the intended interpretation and each kind of symbol-occurrence in the calculus. We seem entitled therefore to presume the mutual consistency of the various principles of inductive syntax that are thus derivable.

Glossary of Technical Terms

The glossary is not intended as a restatement, but as a conveniently abbreviated reminder of the main informal definitions and criteria of use that occur in the text. References to these more detailed definitions and use-criteria are given in the Index. It is assumed that the reader is already familiar with the standard terminology of elementary logic.

Acceptability. A proposition is said to be acceptable at a given time and place, relatively to a given problem, so far as it deserves to be thought the best solution of that problem which is available at that time and place.

Canonical Test. A proposition is said to report the results of a canonical test t_i on a testable hypothesis U if and only if it reports the outcomes of trials of U in each possible combination of the variants of the first i relevant variables (out of the ordered list of variables that are relevant to the particular set of materially similar propositions to which U belongs, preceded by an appropriately selected subset of the variable that is mentioned, or has a variant that is mentioned, in the antecedent of U), where each outcome is said not to be affected by any variant of any other variable than these.

Conjunction Principles for Universal Hypotheses (see also **General Conjunction Principle** and **Instantial Conjunction Principle**). For any E and any universal propositions U_1 and U_2, if $s[U_1] \geqslant s[U_2]$, then $s[U_1 \& U_2] = s[U_2]$, and if $s[U_1, E] \geqslant s[U_2, E]$, then $s[U_1 \& U_2, E] = s[U_2, E]$.

Consequence Principle for Evidential Propositions. For any E, F and H, if F is a consequence of E according to some non-contingent assumptions, such as laws of logic or mathematics, then $s[H, E] \geqslant s[H, F]$.

Consequence Principles for Hypotheses. For any E, I and H, if I is a consequence of H according to some non-contingent assumptions, such as laws of logic or mathematics, then $s[H] \leqslant s[I]$ and $s[H, E] \leqslant s[I, E]$.

Correlational Generalisation. A proposition is said to be a correlational generalisation if and only if it correlates variants of one (qualitative or quantitative) variable with variants of one or more others.

Equivalence Principles. For any E, F, I and H, if E is equivalent to F, and H to I, according to some non-contingent assumptions, such as laws of logic or mathematics, then $s[H] = s[I]$ and $s[H, E] = s[I, F]$.

Evaluation-function. A monadic (dyadic) inductive function is said to be an evaluation-function if and only if it grades the value of what is signified by the antecedent of an evaluative generalised conditional (on the evidence stated by a certain proposition).

Full support. A proposition is said to be fully supported if and only if it is either (i) a proposition that is deducible from non-contingent assumptions (such as laws of logic or mathematics), or (ii) a testable proposition that is not falsified in any trial composing canonical test t_n, where n variables are inductively relevant to it, or (iii) a proposition that is deducible from one or more such testable propositions according to non-contingent assumptions.

Functor. The notation used to signify a function is called a 'functor'.

General Conjunction Principles. For any E, I and H, if $s[H] \geqslant s[I]$, then $s[H \& I] = s[I]$, and if $s[H, E] \geqslant s[I, E]$, then $s[H \& I, E] = s[I, E]$.

Inductive Function. A (monadic or dyadic) function is said to be an inductive function if and only if it satisfies axioms for the principles of inductive syntax.

Inductively Modified Version. See **Inductive Simplicity.**

Inductive Relevance. A (natural, or legal, etc.) variable is said to be inductively relevant to a particular set of materially similar propositions if and only if no expression describing a variant of the variable is, or is definable in terms of, one or more members of the category of non-logical expressions determining that particular set of materially similar propositions, and also either (i) the variable is a two-variant variable and at least once some unmodified member of that set is falsified when one variant of the variable is present and positively instantiated when the second variant is present (in otherwise similar circumstances), or (ii) the variable has more than two variants and each of these is a variant of one or more two-variant variables that are relevant (see also **List of Relevant Variables**). A variable is said to be inductively relevant to a testable hypothesis U if and only if either (i) it is inductively relevant to the particular set of materially similar hypotheses to which U belongs or (ii) it is an appropriately selected subset of a variable v such that either v itself or a variant in this subset is mentioned in the antecedent of U.

Inductive Simplicity. A proposition U′ is said to be an inductively modified version of an unmodified and testable proposition U, or of any of U's equivalents, and to have ith grade inductive simplicity, or to be an ith grade version of U, where $i > 0$, if and only if U′ is like U except for having an appropriate allusion in its antecedent to each of the $n-i$ variables that come last in the ordered list of variables that are inductively relevant to U. A 0th grade version of U is like U except for having as antecedent the conjunction of the antecedent of U with its own denial.

Inductive Syntax. A proposition is said to be a principle of inductive syntax if and only if it states something about what logically implies or is implied by an assessment of inductive support or of inductively permissible simplification, as distinct from what are the empirically corrigible truth criteria for such assessments: e.g., the Equivalence Principles, the Instantial Comparability Principles, or the General Conjunction Principles.

Information-function. A function of two propositions E and H, signified by 'i[E,H]', is said to be an information-function if and only if it measures or grades the extent of the information conveyable by asserting E in answer to the question whether H is true.

Instantial Comparability Principles. For any E, E′, P, P′, U and U′, if U and U′ are first-order generalisations and P and P′ are just substitution-instances of U and U′, respectively, then $s[U] > s[U']$ if and only if $s[P] > s[P']$ and $s[U,E] > s[U',E]$ if and only if $s[P,E] > s[P',E']$.

Instantial Conjunction Principles. For any E, P_1, $P_2, \ldots$, P_n and U, if U is a first-order generalisation and P_1, P_2, ..., and P_n are just some substitution-instances of U, then $s[P_i] = s[P_1 \,\&\, P_2 \,\&\ldots\&\, P_n]$ and $s[P_i, E] = S[P_1 \,\&\, P_2 \,\&\ldots\&\, P_n, E]$. (See also **Conjunction Principle for Universal Hypotheses** and **General Conjunction Principle.**)

List of Relevant Variables. The list of relevant variables for a particular set of materially similar propositions includes all and only the inductively relevant variables for that set, except that it excludes (i) any such variable which is wholly included within another such, and (ii) any variable consisting of two contradictory variants V^1 and not–V^1 if another variable is included which consists of n contrary variants V^1, V^2, . . . V^n; and also, if one variant of a relevant variable is a special case of some variant of another relevant variable, the former should be ignored in favour of the latter.

Materially Similar Proportions. A set of propositions is said to be a particular set of materially similar proportions if and only if it contains all and only (i) the propositions that may be constructed from a given semantical category of non-logical expressions with the aid of standard logical terms and (ii) the inductively modified versions of these.

Natural Variable. See **Variable.**

Necessary Condition. A logical property of an n-adic property or relation R, where $n \geqslant 1$, is said in Chapter VI to be a necessary (logical) condition of R if and only if it is definable without the use of existential quantification (or of the denial of universal quantification) over non-logical variables.

Ratification. A ratification is a justification for some semantical feature or features that are shared by all the natural languages of contemporary civilisation.

Relevant Variable. See **Inductive Relevance.**

Simplification-function. A monadic (dyadic) inductive function is said to be a simplification-function if and only if it grades the inductive simplicity of the simplest fully supported version of a hypothesis (on the evidence stated by a certain proposition).

Simplicity. See **Inductive Simplicity.**

Sufficient Condition. A logical property of an n-adic property or relation R, where $n \geqslant 1$, is said in Chapter VI to be a sufficient (logical) condition of R if and only if R′ has this property only if R′ also belongs to or relates all and only the things that R belongs to or relates.

Support-function. A monadic (dyadic) inductive function is said to be a support-function for a particular set of materially similar propositions if and only if it assesses the extent of the inductive support that exists for any hypothesis belonging to that set by reference to the results of canonical tests (on the evidence stated by a certain proposition).

Test. A proposition is said to report the results of a test on a testable hypothesis U if and only if it reports the outcomes of trials of U in some possible combinations of the variants of one or more variables that are relevant to U. (See also **Canonical Test**.)

Testable Hypothesis. A proposition U is a testable hypothesis if and only if it is a generalised conditional that is seriously open to falsification by one or more variants of the variables that are inductively relevant to the particular set of materially similar propositions to which U belongs.

Trial. A proposition reports the outcome of a trial of a testable hypothesis U if and only if it states either the joint satisfaction of the antecedent and consequent of U, or the satisfaction of the antecedent and the non-satisfaction of the consequent, in some variant of a variable, or in some possible combination of variants of variables, which is or are relevant variables for U.

Uniformity Principles. For any U and P, if U is a first-order generalisation and P is just a substitution-instance of P, then $s[U] = s[P]$ and $s[U, E] = s[P, E]$.

Unmodified Hypothesis. A member of a particular set of materially similar propositions is said to be an unmodified hypothesis if all the non-logical expressions occurring in it are members of the semantical category determining that set.

Variable. A (qualitative or quantitative) (natural, legal, moral, grammatical, or logico-philosophical, etc.) variable is a set of (observationally, juridically, ethically, grammatically, or logico-philosophically, etc.) identifiable kinds, characteristics or circumstances of which the descriptions are either contrary to or contradictory to one another. To avoid confusion with this epistemological sense of 'variable', that word is not used in the present book in its normal formal-logical sense as a type of symbol (except within the expression 'operator-variable' in §§21–22).

Variant. A variant of a (natural, legal, moral, grammatical, or logico-philosophical, etc.) variable is any one of the kinds, characteristics or circumstances that constitute it.

Version. See **Inductive Simplicity.**

Index